Steller Evolution Unveiled

Revealing the Cosmic Drama of Stars from
Birth to Twilight

NAGINDERPAL SINGH

Steller Evolution Unveiled

By Naginderpal Singh

ISBN: 9798866227938 (Paperback)

Cover Design: Naginderpal Singh

Interior Design: Naginderpal Singh

Publisher: The Enlightenment

Published in USA

First Edition: 2023

DEDICATED TO

I am dedicating this book to my parents Sardar Narinder Singh and Late Sardarni Gurjeet Kaur

FOREWORD

In the big theater of the cosmos, there exists a wonderful drama that plays over eons, a narrative of celestial bodies that form the very fabric of our universe. It is a narrative about stars, those dazzling beacons that have caught the imagination of mankind for millennia. "Stellar Evolution Unveiled" leads you on a fascinating trip into the core of stellar worlds, presenting a complete analysis of the life cycles of these cosmic giants.

As we begin on this voyage, we find ourselves in the middle of a cosmic change, from the birth of stars in the dark depths of molecular clouds, to their flaming brightness on the main sequence, and the stunning crescendo of their last acts. Through rigorous study and elegant words, the author walks us through the complicated mechanisms that regulate the growth of stars, exposing the delicate dance of gravity, fusion, and radiation that supports these celestial creatures.

This book does more than explain the individual lives of stars; it uncovers their function as cosmic alchemists, producing components that constitute the fundamental substance of our existence. We see the creation of heavy metals, scattered throughout the universe following stellar death, laying the seeds for future generations of stars and planets. This discovery of stellar nucleosynthesis highlights the fundamental interconnectivity of all celestial entities, underlining that we are, in the purest sense, stardust.

Beyond the individual, the book navigates through the fabric of star clusters and populations, giving a prism through which we might grasp the evolutionary history of galaxies. Through binary star systems, we get insight into the dynamic interaction of stellar partners, highlighting the various ways in which stars alter their cosmic neighbors.

The tale stretches into the worlds of exoplanets and planetary systems, sparking the imagination with the idea of life beyond our home planet. We consider the existence of habitable zones and the tantalizing potential of discovering alien life. This inquiry not only enhances our scientific knowledge but also creates a feeling of awe that transcends the bounds of our reality.

In exploring the future of stars and the universe, the book travels into the hypothetical but enticing territory of sophisticated civilizations and the cosmic energy resources they may exploit. It motivates us to picture a future where mankind, equipped with knowledge and invention, navigates the developing cosmos with wisdom and insight.

"Stellar Evolution Unveiled" is not only a book about stars; it is an invitation to join in the wonder and mystery of the cosmos. It is a monument to the endless curiosity that drives human discovery, and a celebration of the awe-inspiring beauty that pervades the universe. May this trip through the phases of star growth encourage you to look at the night sky with fresh amazement and to explore the unlimited possibility that lies inside the depths of space.

Narinder Singh

PREFACE

In the great expanse of the sky, stars have long been the heavenly lights that capture our imagination and invite us to uncover their secrets. From ancient mythology to cutting-edge astrophysics science, our curiosity with these cosmic beings has remained through the millennia. "Stellar development Unveiled" begins on a voyage to demystify the life cycles of stars, unveiling the subtleties that control their birth, development, and ultimate destiny.

This book is a monument to the constant curiosity of mankind, motivated by an insatiable yearning to grasp the world that cradles us. As we stand on the cusp of an age of extraordinary astronomical discovery, the need for a thorough and accessible understanding of star development has never been higher. With each chapter, we travel deeper into the center of stars, guided by the cumulative knowledge and experience of generations of astronomers and astrophysicists.

The story inside these pages is constructed with the objective of making complicated astrophysical topics approachable to readers of all backgrounds. Through lucid explanations, supplemented by diagrams and real-world examples, we strive to demystify the dynamics that control the lives of stars. Whether you are an aspiring astronomer, a student of astrophysics, or just a curious explorer of the universe, "Stellar Evolution Unveiled" encourages you to embark on a tour through the cosmic tapestry.

While every attempt has been made to offer correct and up-to-date material, the discipline of astrophysics is a dynamic one, continually developing with each new observation and theoretical advance. Therefore, readers are invited to enrich their investigation with the newest research and discoveries in the subject.

Ultimately, this book is a dedication to the stars themselves, those celestial lights that have guided humanity's journey through time and space. It is an invitation to peek into the center of the universe, to see the majesty of star development, and to consider our role within this cosmic symphony.

With eyes focused upwards and hearts open to the glories of the universe, let us start on this adventure together, as we reveal the everlasting mysteries of stellar development.

Naginderpal Singh

ACKNOWLEDGMENTS

I am sincerely thankful for the steadfast support and infinite love shown to me by my parents, Sardar Narinder Singh and Late Sardarni Gurjeet Kaur, during the process of publishing "Stellar Evolution Unveiled". Their support, knowledge, and confidence in my aspirations have been the guiding light that guided my way.

My father, Sardar Narinder Singh, has been a tower of strength and a source of inspiration. His steadfast trust in the power of information and education has formed my worldview and spurred my ambition to dive into the secrets of the universe. His lectures on endurance and determination have been important in this attempt.

Though she is no longer with us, Late Sardarni Gurjeet Kaur's impact continues to echo in every part of my life. Her caring nature, infinite compassion, and enthusiasm for learning have left an unforgettable effect on my character. Her memory is a source of strength that pulls me on, reminding me of the value of enthusiasm and curiosity in the quest of knowledge.

This effort would not have been achieved without the sacrifices and encouragement offered by my parents. Their conviction in my ability and their unfailing support have been the cornerstone of our quest. I dedicate this book to them, with great appreciation and an undying love that has no boundaries.

With heartfelt gratitude,

Naginderpal Singh

CONTENTS

Chapter 1: Introduction to Stars

1.1 The Celestial Tapestry

The notion of the celestial tapestry conjures a vivid vision of the night sky as a fabric embroidered with the threads of stars, each one a unique and bright point of light. It embodies the awe-inspiring beauty and intricacy of the cosmos, enticing us to consider the mysteries that lie beyond our earthly domain. In this chapter, we go on a quest to grasp the complex interaction of stars within this cosmic fabric, diving into their origins, life cycles, and ultimate fates.

The chapter starts by unraveling the birth of stars, a process veiled in the depths of molecular clouds. These massive clumps of gas and dust serve as the cosmic nurseries where stars come into being. Gravitational forces acting upon these clouds commence a dance of matter, culminating to the development of protostars - young celestial entities in their embryonic period. This phase sets the scene for the stellar development that follows, and it is here that the seeds of a star's fate are sowed.

As we advance, we investigate the intriguing anatomy of stars, diving into their composition and interior structure. From the fiery core, where nuclear fusion drives their brightness, to the radiative and convective layers that carry energy outward, stars are shown to be dynamic and developing beings. The complexity of the fusion processes inside their cores, whether via the proton-proton chain or the CNO cycle, expose the mechanisms that preserve a star's brightness and stability throughout its main sequence period.

The Celestial Tapestry, thus, offers as a dramatic introduction to the stellar lifecycle, presenting a vivid image of the cosmic stage upon which stars are born, grow, and finally meet their end. It establishes the groundwork for further research into the various phases of star development that will follow in following chapters. This chapter, by fostering a feeling of awe and admiration for the cosmic marvels above, sets the tone for a complete voyage into the realms of astrophysics, uncovering the mysteries kept inside the heart of each star.

In summation, "The Celestial Tapestry" conveys the essence of the cosmos as a tapestry decorated with the bright gems of stars. It walks the reader through the captivating process of star creation amid molecular clouds and reveals the underlying anatomy of these celestial beings. By presenting this panoramic picture of stellar origins, the chapter awakens interest and establishes the framework for a better knowledge of the complicated phases of stellar development that lie ahead.

1.2 Birth of Stars

The creation of stars is a spectacular cosmic process that develops from the interaction of gravity, gas, and cosmic time scales. Deep inside huge molecular clouds, pockets of dense gas and dust collapse under the unrelenting pull of gravity, commencing the process of star creation. This gravitational collapse happens when the pressure inside a cloud exceeds the outer forces, compressing the center and causing it to heat up. As temperatures increase, the core becomes a protostar, marking the beginning phase in a star's life cycle.

During this protostellar phase, the core's gravitational energy is transformed into heat, leading to the commencement of nuclear fusion. This marks a significant step, as the protostar starts to emit its own light, marking its birth as a real star. The temperature and pressure in the core spark nuclear processes, mostly involving hydrogen atoms fusing together to generate helium. This process releases a vast quantity of energy in the form of radiation, counteracting further gravitational collapse. Consequently, a stable equilibrium is achieved, giving birth to the main sequence phase.

Analysis

The creation of stars encompasses the beauty and intricacy of cosmic phenomena. It is a tribute to the complicated dance of gravitational forces, gas dynamics, and thermodynamics that control the universe. The creation of a star is not an isolated occurrence, but rather a dynamic interaction within the greater setting of molecular clouds, where huge quantities of gas and dust serve as cosmic cradles for embryonic stars.

The gravitational collapse that commences star creation is a process regulated by basic physics laws. It exhibits the unstoppable force of gravity acting on a cloud's mass, a phenomena neatly described by the rules of classical physics. This gravitational collapse, however, must be carefully regulated by the internal pressure of the cloud to avoid rapid collapse, eventually leading to the development of a stable protostar.

The transformation from protostar to real star is a critical phase in this cosmic drama. It is the moment at which nuclear fusion, the lifeblood of stars, ignites. This process, propelled by the tremendous temperatures and pressures in the core, releases a flood of energy, giving birth to the star's brightness. The conversion of hydrogen to helium in the star's core unleashes an astounding amount of energy, establishing the groundwork for a stable life. This equilibrium marks the hallmark of a star's life on the main sequence.

In essence, the creation of stars is a monument to the delicate balance of physical forces at play in the cosmos. It illustrates the harmonic interaction of gravity, gas dynamics, and nuclear physics in the cosmic symphony. The journey from a collapsing cloud to a bright star is a monument to the grandeur and complexity of the natural world, compelling us to explore the fundamental secrets that define our universe.

1.3 Stellar Anatomy

In the enormous fabric of the cosmos, stars are the essential building elements that determine the structure and dynamics of the universe. Chapter 1 of "Stellar development Unveiled" analyzes the detailed anatomy of stars, diving into their composition, structure, and the underlying physical processes that drive their development.

The chapter opens by establishing the scenario, highlighting the role of stars in the big cosmic tale. It offers the groundwork for a fuller knowledge of stellar development by emphasizing the crucial role stars play in the creation of energy, the synthesis of elements, and their final dispersion into space.

Stellar anatomy is not simply a theoretical idea; it is a direct result of the interaction of the four basic forces in the cosmos. This chapter introduces readers to the basic factor guiding a star's existence, gravity. It describes how gravitational collapse leads to the birth of stars from huge molecular clouds. The fusion of hydrogen into helium deep inside a star's core acts as the fundamental engine powering the star, supplying the energy that supports its existence.

To comprehend the structure of stars, readers are directed through the interior layers of a star. Stellar anatomy displays the core, radiative zone, and convective zone, each layer distinguished by specific physical features and behaviors. The chapter underlines the critical function of nuclear fusion at the core, where temperatures and pressures are sufficient to start the fusion of hydrogen, unleashing enormous quantities of energy in the process.

Beyond the core, the radiative zone depicts the area where energy created in the core is transmitted by radiation. Above this layer sits the convective zone, where energy is transmitted by the physical movement of heated material, like the boiling of a cosmic cauldron.

The chapter presents a thorough description of these locations, highlighting the methods of energy transfer, the creation of electromagnetic radiation, and the physical processes responsible for the emission of energy into space.

In summary, Chapter 1 serves as the essential doorway to the research of star development. It not only educates the reader to the underlying anatomy of stars but also instills a feeling of wonder and awe for these cosmic phenomena. Understanding the structure and composition of stars is crucial in revealing the mysteries of their development, which will be the subject of later chapters in this exciting voyage through the universe.

Chapter 2: The Protostellar Phase

2.1 From Molecular Clouds to Protostars

The voyage of a star starts in the chilly, dense center of a molecular cloud, a huge area of space where gas and dust combine under the pull of gravity. These molecular clouds act as cosmic nurseries, encouraging the birth of celestial objects. Within these clouds, pockets of somewhat denser material form, laying the scene for the creation of a protostar. This key period in star development shows the complicated interaction between gravitational collapse, angular momentum, and the production of accretion disks.

As molecular clouds break, gravity exerts its supremacy, bringing together particles of gas and dust. This gravitational collapse commences a sequence of events that lead to the development of a protostar. The process is analogous to nature's own sculptor, methodically creating the core of what will ultimately become a star. The force of gravity compresses the material, creating heat, and as the core shrinks, its temperature increases. This spike in temperature signals the fledgling light of a protostar, a prelude to the brilliant brightness it will one day acquire.

Angular momentum, an intrinsic feature of the collapsing material, plays a critical role in structuring the forming protostar. As the cloud decreases, its rotational speed rises, saving angular momentum. This event leads to the creation of a flattened, spinning accretion disk encircling the protostar. The disk works as a store of substance, giving the expanding star with the nutrition it needed to mature. The interaction between gravitational forces and angular momentum produces a vivid depiction of the delicate dance inside molecular clouds, revealing the beauty of cosmic processes on a large scale.

The creation of accretion disks surrounding protostars signals a critical step in their growth. These disks not only support the evolution of the core protostar but also offer the birthplace for prospective planetary systems. Within the limits of the disk, dust grains collide and merge, generating planetesimals that may someday consolidate into planets. This process of planetary formation is an intrinsic component of the greater story of cosmic development, where stars and their prospective planetary partners are born from the same celestial cradle.

In conclusion, the journey from molecular clouds to protostars is a captivating narrative of cosmic genesis. Within the center of molecular clouds, gravity orchestrates the development of a protostar, laying the scene for the emergence of a new stellar entity. The interaction between gravitational collapse, angular momentum, and the development of accretion disks dictates the fate of these celestial objects. The creation of accretion disks not only nourishes the expanding protostar but also supplies the origin for prospective planetary systems. This phase of star development highlights the complicated dance of forces and processes that control the universe, reminding us of the incredible beauty and complexity that exists beyond our earthly habitat.

2.2 Gravitational Collapse and Accretion Disks

Gravitational collapse is a critical step in stellar development, signifying the shift from the earliest phase of a protostar to a mature, energy-producing star. At its root, this phenomena rests on the interaction between gravitational forces and opposing pressure gradients. As a molecular cloud fragment gravitationally collapses, it experiences a process of compression, generating a rise in temperature and pressure inside the center. When a certain density and temperature are achieved, nuclear fusion is triggered, marking the commencement of the star's main-sequence phase.

One key feature of this shift is the creation of accretion disks. These disks are flat, revolving formations made of gas, dust, and other debris that ring the core object. In the context of star formation, they derive from the conservation of angular momentum, a concept that mandates that a collapsing cloud must preserve its original angular momentum. As the cloud shrinks, it starts to spin faster, finally flattening into a disk. This disk serves as the reservoir of matter from which the core protostar accretes mass, promoting its expansion.

Accretion disks have a key function in controlling the mass accretion rate onto the core protostar. The disk's material spirals inward owing to viscous forces inside the disk itself. These forces occur from interactions between neighboring rings of material, which transport angular momentum outward and mass inward. Consequently, matter increasingly migrates towards the center, where it nourishes the expanding protostar. This mechanism achieves a delicate balance between inward accretion and outward transmission of angular momentum, guaranteeing a continual supply of material for the core star's evolution.

The study of accretion disks goes beyond the field of star formation and is also useful in scenarios like double star systems and active galactic nuclei. In binary systems, an accretion disk develops around a compact object, such as a white dwarf, neutron star, or black hole, as it sucks material from its partner. This process may rise to phenomena like X-ray binary emissions and novae. Moreover, in the case of active galactic nuclei, supermassive black holes at the heart of galaxies are surrounded by huge accretion disks, emitting strong radiation and showing properties found in quasars and blazars.

In summary, gravitational collapse and the subsequent creation of accretion disks are critical phases in the life cycle of stars. They symbolize the shift from a cloud of gas and dust to a fledgling star, powered by the relentless force of gravity. Accretion disks not only support the formation of the core protostar but also play key roles in a number of astronomical phenomena, ranging from double star systems to the most energetic occurrences in the universe. Understanding these processes is vital for deciphering the complexity of star development and the larger cosmic world.

2.3 Protostellar Evolutionary Tracks

Protostellar evolutionary tracks follow the transforming path of a nascent stellar object from its origin inside a molecular cloud to its final emergence as a full-fledged star on the main sequence. This important period in a star's life cycle involves a variety of dynamic processes that form its ultimate traits and attributes.

At the start, a protostar develops from the core of a thick molecular cloud, a huge conglomeration of gas and dust. Gravitational collapse initiates this inception, when the self-gravity of the cloud surpasses the pressure forces counteracting it. As the falling material accrues onto the protostar, it produces a revolving accretion disk. This disk acts as the feeding tube, pouring material onto the newborn star while also giving birth to future planetary systems.

The protostellar phase involves a dynamic interaction of gravitational forces, accretion processes, and the complex dynamics of the surrounding environment. This dynamic emerges in several sorts of protostars, such as Class 0, Class I, and Class II, each reflecting unique phases of development. Class 0 protostars are the youngest, remaining completely immersed inside their natal cocoon of dust and gas. They are characterized by tremendous accretion rates and intense discharges of material. As the protostar advances to Class I, it starts to clean its immediate environs, exposing its fundamental core. Class II protostars, also known as pre-main sequence stars, have lost most of their surrounding material and are well on their road to reaching the main sequence.

The evolutionary trajectories of protostars are not uniform; they rely strongly on beginning variables like as mass, angular momentum, and ambient density. Higher mass protostars may have a more tumultuous and fast development, while lower mass protostars follow a more steady route. Additionally, the slope of the accretion disk and the angle of outflows impact the visible properties of a protostar.

Furthermore, the interplay between the protostar and its surroundings plays a significant part in deciding its eventual destiny. Feedback processes, such the ejection of surplus energy via jets and outflows, may modify the accretion process and impact the development of companion objects like brown dwarfs or planets.

In summary, protostellar evolutionary tracks are a monument to the complicated interaction of physical processes that determine the early stages of a star's existence. The journey from a fledgling core inside a molecular cloud to a detectable protostar on the verge of main sequence integration is distinguished by a number of dynamic stages, each leaving a distinctive impression on the emerging item. Understanding these tracks is crucial to understanding the variety of stars and the possibilities for planetary systems inside our galaxy.

Chapter 3: The Main Sequence

3.1 Fusion Ignition and Stellar Stability

Fusion ignition stands as the cosmic ignition key that changes a dense, gravitationally confined cloud of hydrogen into a star. This critical process happens at the core of a protostar, where tremendous pressures and temperatures encourage nuclear fusion. The major fusion mechanism in stars like our sun is the proton-proton chain reaction, where hydrogen nuclei (protons) fuse to make helium nuclei, freeing a torrent of energy in the process. This release of energy counterbalances the unrelenting inward pull of gravity, maintaining a delicate equilibrium that determines a star's stable phase.

At the core, pressures surge to staggering heights, reaching millions of atmospheres, as the density deepens. These parameters are needed for the commencement of nuclear fusion. Only when hydrogen atoms are crushed with enough force can they overcome their inherent electrostatic repulsion and combine. The core temperature, frequently in the millions of Kelvin, supplies the kinetic energy required to induce these collisions.

However, sustaining excellent stability isn't an easy effort. The rate at which nuclear reactions occur must properly balance with the energy radiated out from the core. This equilibrium is finely choreographed by the interaction between temperature, pressure, and energy transfer systems, mainly convection and radiation. For instance, at the center of our sun, photons of high-energy gamma radiation are constantly produced by nuclear processes, yet they remain confined, dispersing amongst charged particles. This conflict between production and escape maintains the star's balance.

The stability of a star rests on a delicate interaction between the outward pressure from nuclear fusion and the inexorable inward pull of gravity. If the equilibrium is upset, a star may develop or move to a new phase. For instance, when a star exhausts its core hydrogen fuel, the core compresses and temperatures climb. This leads to the enlargement of the outer layers, indicating the shift to the red giant phase. In this phase, helium fusion in a shell surrounding the core further impacts the star's growth.

In summary, the principle of fusion ignition and stellar stability supports the fundamental existence of stars. It displays a delicate balance between the inward pull of gravity and the outward pressure created by nuclear fusion. This mechanism, relying on exact circumstances of temperature and pressure, guarantees that a star stays in a stable state for a large amount of its existence. Any disturbances to this equilibrium, either by changes in the supply of fuel or adjustments in the fusion process, lead to major shifts in the star's growth, eventually dictating its destiny. Understanding this delicate dance of forces inside stars not only increases our grasp of their lifespan but also gives a window into the bigger mechanisms driving the universe.

3.2 Hydrostatic Equilibrium and Radiative Zones

Hydrostatic equilibrium and radiative zones are key ideas in understanding the interior structure and behavior of stars. Hydrostatic equilibrium refers to the balance between the inward force of gravity and the outward pressure produced by the heated gasses inside a star. This balance is necessary for a star to retain its stable condition, preventing it from collapsing under its own gravitational force or growing wildly. In basic words, it guarantees that the star maintains in a condition of relative equilibrium, allowing for the continued nuclear fusion reactions that fuel it.

Within a star, there are separate regions called as radiative zones. These zones are defined by the major route of energy transmission, which is by the emission of photons. In a radiative zone, energy is transferred outward by high-energy photons, created by nuclear processes in the core. This energy then diffuses outwards, experiencing a complicated process of absorption and re-emission by the thick plasma of the star's core. Radiative zones are often located in the core of stars, where temperatures and pressures are very high.

To assess the relevance of hydrostatic equilibrium and radiative zones, it is necessary to grasp their interconnection. Hydrostatic equilibrium is the cornerstone of a star's stability, acting as the driving force behind the star's capacity to retain its structure and balance despite gravitational forces. Without this balance, stars would either collapse under their own weight or grow uncontrollably, leaving them incapable of sustaining nuclear fusion.

Radiative zones serve a vital part in the energy transmission process inside a star. In the core, where temperatures reach millions of degrees, nuclear fusion reactions occur, producing a huge quantity of energy. This energy is initially in the form of high-energy photons, which must travel through the radiative zone to reach the outer layers of the star. The process of absorption and re-emission inside the radiative zone guarantees that energy is successfully carried to the star's surface, where it is finally radiated out into space as light and heat.

Moreover, the presence of radiative zones gives vital information regarding a star's developmental stage. For instance, during the main sequence phase, the core of a star is largely a radiative zone, where hydrogen nuclei fuse to generate helium. As a star matures, and its core's temperature and pressure vary, the borders of the radiative zone may move. This may lead to changes in a star's brilliance, size, and final destiny.

In summary, hydrostatic equilibrium and radiative zones are basic ideas that regulate the behavior and development of stars. Hydrostatic equilibrium establishes a precise balance between gravitational forces and internal pressure, enabling a star to retain its stability. Radiative zones, on the other hand, serve a key function in moving energy from the core to the outer layers of the star, impacting its brightness and development. Understanding these principles gives a significant insight into the life cycles and behaviors of stars, leading to our greater grasp of the cosmos.

3.3 The Proton-Proton Chain and CNO Cycle

The Proton-Proton Chain and the Carbon-Nitrogen-Oxygen (CNO) Cycle are two essential nuclear processes that fuel the cores of stars, including our Sun. They are the principal methods via which hydrogen nuclei (protons) fuse together to generate helium, releasing tremendous quantities of energy in the process. These activities play a vital role in determining a star's energy output and developmental course.

The Proton-Proton Chain, which prevails in stars like our Sun, includes a succession of nuclear events. In the first stage, two protons smash and generate a deuterium nucleus, a positron, and a neutrino. This is a difficult operation, necessitating that the protons overcome the electrostatic repulsion between them by the high temperatures and pressures in the star core. Subsequently, another proton collides with the deuterium nucleus, leading to the creation of helium-3 and the discharge of a gamma-ray photon. Finally, two helium-3 nuclei unite to generate a helium-4 nucleus, unleashing two protons in the process. This cycle finally turns four protons into a single helium-4 nucleus, coupled with the release of a large quantity of energy in the form of light and heat.

In contrast, the CNO Cycle is more prominent in big stars, particularly those with more than 1.3 times the mass of the Sun. This process depends on carbon, nitrogen, and oxygen isotopes as catalysts. In the CNO Cycle, carbon-12 acts as a catalyst, allowing hydrogen nuclei to be transformed into helium-4. The cycle comprises numerous phases, including the capture of a proton by carbon-12, which leads to the creation of nitrogen-13, followed by the emission of a positron. Subsequent reactions result in the conversion of nitrogen-13 into carbon-12, and the cycle continues. While the CNO Cycle is less effective than the Proton-Proton Chain in turning hydrogen into helium, it becomes more significant in more massive stars with greater temperatures.

The relevance of these processes rests not just in their position as energy sources but also in their influence on a star's evolutionary trajectory. Stars spend the most of their lifetimes on the main sequence, where hydrogen fusion predominates. Understanding the complexity of the Proton-Proton Chain and CNO Cycle enables astrophysicists to calculate the lifespan of stars, anticipate their eventual destinies, and examine the synthesis of elements inside their cores.

In summary, the Proton-Proton Chain and CNO Cycle reflect the cornerstone processes by which stars create energy. The Proton-Proton Chain, prominent in stars like our Sun, includes a chain of events turning protons into helium-4. The CNO Cycle, on the other hand, is more prominent in big stars and depends on carbon, nitrogen, and oxygen isotopes as catalysts for hydrogen fusion. These mechanisms not only maintain a star's brilliance but also impact its evolutionary course, highlighting their basic relevance in astrophysics.

Chapter 4: The Hertzsprung-Russell Diagram

4.1 Spectral Classification

Spectral classification is a basic method in the study of stars, allowing astronomers to identify and comprehend a vast variety of celestial objects based on the qualities of their emitted light. This method, first created by Annie Jump Cannon and revised by others, arranges stars into discrete spectral classes marked by letters, most notably the O, B, A, F, G, K, M series. Each letter corresponds to a certain temperature range, with O-type stars being the hottest and M-type stars the coolest. This categorization method gives vital information regarding a star's physical features, evolutionary stage, and even its potential habitability.

O-type stars, typified by their bluish-white hue and surface temperatures upwards of 30,000 Kelvin, are the most luminous and massive stellar entities. Due to their high energy output, O-type stars are generally short-lived, consuming their nuclear fuel within a few million years. This leads to their commonly being discovered in areas of intensive star formation, such as inside huge star clusters and the spiral arms of galaxies. Additionally, these stars are vital in changing their surrounding environs via intense stellar winds and, ultimately, supernova explosions.

In contrast, M-type stars comprise the most frequent stellar class in the cosmos. Their reddish tint and temperatures ranging from 2,400 to 3,600 Kelvin are suggestive of their reduced mass and luminosity compared to O-type stars. While M-dwarfs may not be as physically beautiful as their hotter counterparts, their lifespan far exceeds them. Some M-dwarfs have been detected to burn consistently for billions of years, possibly offering stable settings for planetary systems to emerge.

The spectrum categorization scheme also uncovers crucial information about a star's chemical makeup. Spectral lines, which show as dark or brilliant lines in a star's spectrum, reflect the existence of various elements in the star's outer layers. Elements like hydrogen, helium, calcium, and iron leave characteristic marks in the spectrum, enabling astronomers to identify the elemental composition of a star. This data is vital for understanding the processes of nucleosynthesis inside stars and the subsequent dispersion of elements across the universe.

Moreover, spectral categorization plays a key role in the hunt for alien life. By detecting stars with comparable spectral properties to our Sun, astronomers may locate prospective solar analogs. These stars are of special importance in the hunt for habitable exoplanets, since they offer a steady energy supply for possible life-bearing worlds. Additionally, spectral analysis may indicate the existence of components like oxygen and water vapor in the atmospheres of exoplanets, further narrowing the hunt for habitable zones outside our solar system.

In conclusion, spectral classification is a vital technique in contemporary astronomy, helping scientists to classify and comprehend the tremendous variety of stars in the cosmos. Through this categorization system, astronomers obtain information into a star's temperature, mass, brightness, chemical makeup, and evolutionary stage. This knowledge is not only vital for furthering our understanding of star dynamics but also for the hunt for life outside our solar system. Spectral categorization serves as a tribute to the ability of observation and investigation in understanding the secrets of the universe.

4.2 Luminosity Classes

In the field of stellar classification, luminosity classes play a vital role in classifying stars based on their intrinsic brightness, a quality that is different from their apparent brightness as viewed from Earth. Luminosity classes are designated by Roman numerals, spanning from I to V, with each number reflecting a distinct stage in a star's evolutionary path.

Class I: Supergiants

The first luminosity class, indicated as I, contains the most massive and bright stars in the cosmos. These behemoths, typically hundreds of times more massive than the Sun, emit an immense amount of energy. Despite their extraordinary brilliance, they are rather uncommon in our galaxy, with noteworthy examples being Betelgeuse and Rigel in the constellation Orion. Their huge size and energy production make them essential in changing the dynamics of their host galaxies.

Class II: Bright Giants

Class II stars, designated as luminous giants, are likewise significantly brilliant but less massive than supergiants. They represent an intermediate stage in a star's development, generally occurring after it has exhausted its core hydrogen fuel. Examples of brilliant giants include Aldebaran in the Taurus constellation. While not as energetically powerful as supergiants, brilliant giants nonetheless wield enormous impact over the surroundings they inhabit.

Class III: Giants

Giant stars, defined as class III, are stars that have developed beyond the main-sequence phase, having exhausted their core hydrogen. These stars have grown and cooled, resulting in a dramatic rise in size. Stars like Arcturus in the Bootes constellation come into this group. Giants are bigger and brighter than main-sequence stars of same spectral type, frequently having a great effect on the areas of space they inhabit.

Class IV: Subgiants

Subgiants, classified by class IV, are transitional stars between main-sequence and giant stars. They display traits of both phases, exhibiting a brightness somewhat greater than that of main-sequence stars of comparable spectral type. As a subgiant matures, it will ultimately grow into a giant. While they may not be as remarkable as supergiants or giants, subgiants play a crucial role in our knowledge of stellar development.

Class V: Main Sequence

The fifth and final luminosity class, class V, encompasses the great majority of stars, including our Sun. Main-sequence stars are at the prime of their hydrogen-burning phase, when nuclear fusion occurs in their cores, turning hydrogen into helium. They demonstrate a steady balance between the inward pull of gravity and the outward push of radiation. These stars offer the basis for our knowledge of stellar development, since they represent the most frequent phase in a star's life cycle.

Analysis:

The luminosity classes give a hierarchical framework for comprehending the broad assortment of stars that inhabit our galaxy and the distant universe. They enable astronomers to classify stars based on their inherent brightness, contributing in the construction of models for stellar evolution. This categorization method also acts as a significant tool in numerous aspects of astronomy, including the study of galactic dynamics, the behavior of star clusters, and the research of stellar populations.

Furthermore, luminosity classes give information into the greater astrophysical background. Supergiants, for instance, play a vital role in the synthesis of heavy metals, which are ultimately spread into the interstellar medium via catastrophic supernova occurrences. Giants and luminous giants, on the other hand, contribute to the enrichment of their host galaxies with materials needed for the development of planets and life as we know it.

In summary, the notion of luminosity classes is a cornerstone of stellar astrophysics. It offers a strong tool for astronomers to classify and analyze the variety of stars, affording a glimpse into the dynamic processes that regulate the development and life cycles of celestial objects.

4.3 Evolutionary Paths on the HR Diagram

The Hertzsprung-Russell (HR) Diagram is a crucial tool in the study of star evolution. It classifies stars based on their brightness and temperature, offering a visual picture of their life cycles. One key component depicted on this map is the evolutionary trajectories that stars pursue during their lives. These pathways are governed by the interplay between a star's starting mass, its internal activities, and outward interactions with its surroundings.

At the outset of their history, stars are positioned on the main sequence, a band that stretches diagonally from the top left (high brightness, high temperature) to the lower right (low luminosity, low temperature) of the HR Diagram. The main sequence is inhabited with stars engaging in hydrogen fusion in their centers. This period may continue for millions to billions of years, depending on the star's initial mass. Lower-mass stars like red dwarfs follow a slow and steady journey, whereas more massive stars travel more swiftly.

As a star exhausts its hydrogen fuel in the core, it embarks on a new trajectory on the HR Diagram. For stars of lesser mass, such as our Sun, this transition is marked by an expansion and cooling phase, culminating in the star becoming a red giant. This movement to the upper-right part of the picture is a characteristic of a star's progression away from the main sequence. Here, helium fusion originates in the core, while hydrogen fusion occurs in a surrounding shell.

In contrast, high-mass stars, those larger than about eight times the mass of the Sun, pursue a more dramatic evolutionary course. After exiting the main sequence, they undergo a quick metamorphosis. Their cores continue to compress and heat until they reach the necessary temperature for helium fusion, resulting to a rapid rise in brightness and a movement towards the top left of the HR Diagram. This phase, known as the helium flash, is a defining period in the life of a high-mass star.

As stars further grow, their courses differ depending on their starting mass. Low-mass stars, like the Sun, will ultimately shed their outer layers in a planetary nebula, leaving behind a hot, dense core known as a white dwarf. These relics eventually cool and fade away over billions of years. Meanwhile, high-mass stars continue a sequence of nuclear events, synthesizing heavier elements until they reach iron, which cannot be further fused. This leads to a catastrophic collapse and, depending on the mass, either a supernova explosion or the birth of a black hole.

In summary, the HR Diagram provides as a visual narrative of a star's journey through its many evolutionary stages. By observing the locations of stars on this diagram, astronomers may determine their present stage and anticipate their future destiny. It is a monument to the interdependence of a star's mass, composition, and internal processes that eventually form the magnificent celestial displays and mysterious relics that inhabit our universe. Understanding these evolutionary routes is vital not only for grasping the life cycles of individual stars but also for decoding the bigger story of cosmic development.

Chapter 5: Red Giants and Supergiants

5.1 The Expansion Phase

The Expansion Phase, a major chapter in the investigation of stellar evolution, illustrates the transforming journey of stars beyond the main sequence. As stars exhaust their hydrogen fuel in the core, they experience a sequence of structural and chemical changes, eventually leading to their growth into red giants and supergiants. This phase shows the complicated interplay of internal mechanisms that dictates the fate of stars.

The Expansion Phase occurs when a star's hydrogen core is exhausted, resulting to the contraction of the helium core. As the core compresses, gravitational energy is turned into thermal energy, causing the outer envelope of the star to expand. This expansion leads to a fundamental alteration in the star's attributes, including its size, temperature, and brightness.

The most noticeable aspect of the Expansion Phase is the ballooning of the star. Red giants and supergiants are sometimes hundreds to thousands of times bigger in radius than they were on the main sequence. This tremendous growth is caused by the growing pressure in the core, resulting from helium burning, notably via the triple-alpha process. The increasing core pressure pulls the outer layers outward, causing the star to bloat.

During this period, stars demonstrate an unusual link between their surface temperature and brightness. Although they have greater radii and colder surface temperatures compared to their main sequence counterparts, red giants and supergiants are substantially more brilliant. This seeming contradiction derives from the large surface area of these big stars, which compensates for their lower surface temperatures.

The Expansion Phase is also indicated by the commencement of helium burning in the core, leading to the production of heavier elements by nuclear fusion. This process is responsible for the synthesis of elements beyond helium, such as carbon and oxygen. Red giants and supergiants become chemical factories, enriching the surrounding universe with these freshly created components.

Throughout the chapter, we study the internal processes driving the Expansion Phase. Stellar pulsations become widespread in red giants, leading these stars to display variations in their brightness. Such pulsations are a result of the star's complicated internal structure, where layers of various compositions and physical circumstances interact. These pulsations give significant insights into the intrinsic features of stars and provide a window into their development.

The Expansion Phase is a vital stage in the life cycle of stars. It lays the groundwork for the ultimate ejection of star material into space, filling the universe with heavy elements. Moreover, the destiny of these enlarged stars, whether they collapse into white dwarfs or suffer spectacular supernova explosions, is controlled by the complicated processes happening during this era. Understanding the Expansion Phase is vital for understanding the range of cosmic occurrences, from the birth of new stars to the formation of heavy elements that define the universe's chemical makeup.

5.2 Helium Shell Burning

Helium shell burning is a vital step in the life cycle of intermediate-mass and massive stars. It happens in the last phases of a star's development, following the depletion of hydrogen in its core. This phase plays a critical role in defining the star's destiny, leading to the production of heavier elements, and in certain circumstances, causing spectacular occurrences like stellar pulsations, shell flashes, and even the final supernova explosion.

During the main sequence period of a star's existence, nuclear fusion predominantly includes the conversion of hydrogen into helium in the star's core. As the star exhausts its hydrogen fuel, the core compresses and heats up, while the outer envelope expands. The increasing temperature and pressure at the core enable helium nuclei (alpha particles) to collide and fuse together, generating heavier elements like carbon, oxygen, and neon via a process known as helium shell burning. This is a vital period for the creation of elements beyond helium in a star's core.

The process of helium shell burning is defined by a sequence of thermal pulses in which the helium shell burns occasionally. These pulses are caused by the unstable behavior of helium fusion in a shell surrounding the inert carbon-oxygen core. When the temperature and pressure in the helium shell grow sufficiently high, helium ignition occurs, resulting to a fast release of energy. This energy pushes the outer envelope of the star outward, causing it to expand and cool, and providing a visual representation of these events in the form of a pulsation.

Helium shell burning has dramatic consequences on the star's development. It may lead to the production of intense stellar winds that spew matter from the star's outer layers. This mass loss may drastically affect the star's composition and brightness, ultimately leading to the development of planetary nebulae or the manufacturing of heavy elements in the outer layers of the star.

In rare situations, helium shell burning may potentially propel the star towards a catastrophic termination. In stars with initial masses between roughly 8 and 10 solar masses, helium shell burning may precipitate a core collapse and subsequent supernova explosion. This signifies the end of the star's existence and the spread of heavy elements into space, enriching the interstellar medium and contributing to the development of future generations of stars and planets.

In conclusion, helium shell burning represents a crucial phase in the lives of intermediate-mass and massive stars. It is responsible for the creation of heavier elements, the onset of thermal pulses, and the potential for dramatic events in a star's history. Understanding the dynamics of helium shell burning is vital for deciphering the intricate mechanisms that govern the life cycles of stars and the universe's chemical history.

5.3 Stellar Pulsations and Variable Stars

Stellar pulsations, a phenomena at the core of Chapter 5, "Stellar Pulsations and Variable Stars," are an intriguing element of stellar development. This chapter digs into the unique patterns displayed by particular stars, giving a fascinating view into their internal dynamics and life cycles.

The chapter opens with an analysis of the triggers and processes driving stellar pulsations. It elucidates how the balance between pressure, gravity, and energy creation inside a star may lead to periodic expansions and contractions. This pulsing phenomena, analogous to a celestial pulse, emerges in numerous sorts of stars, from red giants to Cepheid variables.

A substantial amount of the chapter is dedicated to Cepheid variables, a type of pulsing stars noted for their amazing regularity. These stars act as cosmic yardsticks, allowing scientists to quantify astronomical distances with unparalleled precision. The chapter emphasizes how the luminosity-period connection established by Henrietta Swan Leavitt transformed our view of the universe, establishing the groundwork for Edwin Hubble's revolutionary findings.

Furthermore, the chapter dives into other kinds of variable stars, such as RR Lyrae and Mira variables. It elucidates their distinctive pulsation properties, offering light on their developmental phases and the crucial insights they give into the greater star population of galaxies.

The intellectual complexity of this chapter goes beyond the observable domain. It dives into the theoretical theories and mathematical frameworks applied to interpret star pulsations. The subject spans the use of light curves, period-luminosity relations, and mode identification methods, revealing how astronomers harvest vital information about star interiors from surface measurements.

Moreover, the chapter deals into the larger ramifications of star pulsations. It underlines how these rhythmic oscillations have far-reaching repercussions on a star's general history, impacting its mass loss, chemical enrichment of the interstellar medium, and, eventually, its destiny. This section underlines the complicated connection between pulsations and the larger cosmic environment.

In conclusion, "Stellar Pulsations and Variable Stars" is a vital chapter that not only uncovers the hypnotic rhythms intrinsic to some stars but also gives a full knowledge of their underlying mechanics. It elegantly links together observational data, theoretical models, and their larger astrophysical implications, providing readers a complete perspective of this intriguing part of star development. Through its simple style and intelligent research, this chapter illustrates the immense influence that pulsations have on the cosmos, underlining their crucial position in the magnificent fabric of the universe.

Chapter 6: Stellar Nucleosynthesis

6.1 Fusion Beyond Hydrogen

Nuclear fusion, the mechanism that fuels the stars, is not restricted to the simple union of hydrogen nuclei. Deeper into the cores of stars, where temperatures and pressures reach astounding extremes, helium nuclei, or alpha particles, become the next performers in this cosmic drama. This shift represents a pivotal period in stellar development, as stars boost their energy output and prepare the way for the synthesis of heavier elements.

As a star advances through its evolutionary phases, the hydrogen fuel in its core depletes. The core compresses and warms up, bringing up the temperature to millions of degrees. At this time, hydrogen fusion proceeds in a shell surrounding the core, while helium accumulates in the core itself. When the core temperature hits roughly 100 million degrees, helium fusion starts. Unlike hydrogen fusion, which requires the combining of two protons, helium fusion creates a series of processes known as the triple-alpha process. This process requires the fusion of three helium nuclei to generate carbon.

The triple-alpha process is a complex cosmic dance that relies on the precise balance between the strong nuclear force and the electromagnetic force. The strong nuclear force binds protons and neutrons inside atomic nuclei, whereas the electromagnetic force, which includes the repulsion between positively charged protons, opposes the compression required for fusion. For three helium nuclei to fuse into carbon, the repulsive electromagnetic force must be overcome. This only occurs at the tremendous temperatures and pressures seen in the core of a mature star.

The commencement of helium fusion indicates a key transition in a star's life cycle. The energy emitted by helium fusion greatly surpasses that of hydrogen fusion. This spike of energy leads the outer layers of the star to expand, and the star begins a new period of its existence known as the red giant phase. This expansion eventually leads to the evacuation of the star's outer layers in a stunning phenomenon termed a planetary nebula. What left behind is the exposed core, which, depending on its mass, may further grow into a white dwarf, neutron star, or even a black hole.

The impact of the triple-alpha process goes beyond individual stars. It serves as the genesis of carbon, the primary building component of life as we know it. Carbon-rich material expelled from dying stars ultimately interacts with other elements in space, generating new generations of stars and planetary systems. This process underlines the tremendous interconnection of the universe, revealing how the development of stars influences the very nature of life on Earth.

In summary, the fusion of helium via the triple-alpha process is a critical event in the life of a star. It signifies a shift from hydrogen-based fusion to the synthesis of heavier elements, resulting to the final dispersion of star material that enriches the universe. This process not only feeds the energy production of stars but also plays a critical part in the formation of carbon, an essential ingredient for life. The triple-alpha process serves as a witness to the complicated interaction of nuclear forces that regulate the workings of the cosmos.

6.2 The Stellar Forge: Creation of Elements

Within the scorching crucible of stars, a cosmic alchemy unfolds, giving birth to the varied variety of elements that constitute the very fabric of our universe. This process, known as nucleosynthesis, is the cosmic forge where hydrogen and helium, the primordial building blocks, are turned into heavier elements via the unrelenting fusion events that occur at star cores. It is in this burning furnace that the seeds of life, the carbon, oxygen, and many other elements, are created.

The trip starts at the center of a star, where high pressure and temperatures exceeding millions of degrees Kelvin compel hydrogen nuclei to smash and fuse, producing helium. This mechanism, known as the proton-proton chain, is the engine that fuels main-sequence stars, including our own Sun. As a star grows, it undergoes multiple phases of fusion, where helium nuclei fuse to generate carbon, oxygen, and trace quantities of other light elements. This early phase of nucleosynthesis, dubbed stellar hydrogen and helium burning, lays the groundwork for the formation of heavier elements.

As a star matures and nears the conclusion of its life, a dramatic alteration happens. In the crucible of its core, helium fusion gives birth to the synthesis of heavier elements via processes like the triple-alpha process, where helium nuclei unite to make carbon. Subsequently, in more massive stars, the core temperature and pressure increase, permitting the fusion of carbon and oxygen into even heavier elements like neon, magnesium, and silicon. These phases depict the zenith of the star forge's creativity, when nuclei dance and join, generating components that will ultimately be spread over the universe.

The catastrophic culmination of a large star's existence is characterized by a dazzling supernova explosion, an event of inconceivable intensity that for a short while outshines whole galaxies. Within this massive explosion, elements beyond iron, including gold, silver, and uranium, are produced in the last throes of nucleosynthesis. This supernova forge, where tremendous temperatures and pressures rule, drives these freshly produced elements into space, dispersing them throughout the universe, ready to be absorbed into new generations of stars, planets, and, eventually, life.

The consequences of this star forge are enormous. It is the engine that drives the development of galaxies, the crucible that births the ingredients essential for planetary systems, and the source of the elements that make life as we know it conceivable. Our fundamental existence, from the carbon in our DNA to the oxygen we breathe, owes its birth to the alchemical processes that occur inside the hearts of stars. Understanding this stellar forge not only reveals the formation of elements but also gives a look into the rich fabric of cosmic development that creates our cosmos.

6.3 Supernova Explosions and Heavy Elements

Supernova explosions represent the dramatic finale of a large star's life cycle, releasing immense quantities of energy and spawning some of the most perplexing phenomena in the cosmos. These catastrophic occurrences occur when a star's core can no longer resist gravitational forces, resulting to a fast collapse followed by a massive explosion. The energy generated during a supernova outshines a whole galaxy, temporarily dazzling the universe. One of the most important repercussions of these explosions is the synthesis of heavy atoms, elements beyond helium, which are necessary for the development of planets, life, and eventually, the variety of the cosmos.

As the core of a big star near the conclusion of its evolutionary trajectory, nuclear fusion processes halt, leaving behind a core comprised mostly of iron. Iron is unusual in that it is the most closely bonded nucleus in terms of energy per nucleon. This implies that fusing iron nuclei together needs more energy than it releases, making it the death knell for a falling star's core. In reaction, the core experiences a fast gravitational collapse, with protons and electrons combining to generate neutrons in a process known as neutronization. This collapse is so sudden that it creates a shockwave that passes through the star, eventually leading to the ejection of its outer layers.

The enormous outpouring of energy during a supernova is a spectacular occurrence, one that may temporarily outshine whole galaxies. This brightness occurs from the conversion of gravitational potential energy into kinetic energy, when the core's collapse compresses it to densities rivaling that of atomic nuclei. Subsequently, the rebounding shockwave hurls the outer layers into space at extraordinary speeds, causing a spectacular explosion of light and radiation.

In one moment of celestial grandeur, a supernova may produce as much energy as our Sun would during its whole career.

However, the ultimate cosmic spectacle of supernovae rests in their function as cosmic alchemists. Elements heavier than iron, such as gold, silver, and uranium, owe their existence to the extreme circumstances of a supernova explosion. The tremendous temperatures and pressures created during the explosion allow nucleosynthesis, the forging of heavier elements from lighter ones. Neutrons, created in abundance during the collapse, become absorbed by existing nuclei, catalyzing fast neutron capture processes, or the s-process, culminating in the synthesis of a rich tapestry of heavy elements. These freshly synthesized elements are subsequently expelled into space, spreading across galaxies, ready to be absorbed into the next generation of stars, planets, and, eventually, life.

In conclusion, supernova explosions are the cosmic grand finales of huge stars, unleashing inconceivable quantities of energy and commencing the production of heavy metals. These awe-inspiring events are important in creating the universe, affecting the makeup of galaxies and laying the road for the birth of life. The enormous effect of supernovae highlights their relevance in unraveling the rich fabric of star history and the evolution of the cosmos itself.

Chapter 7: Stellar Death: White Dwarfs

7.1 The Fate of Low-Mass Stars

Low-mass stars, those with a mass less than around 1.5 times that of our Sun, follow a distinct evolutionary path compared to their more massive counterparts. These stars form the majority of stars in the cosmos, and their destiny is a topic of significant interest in astronomy. As a low-mass star advances through its life cycle, it experiences a sequence of modifications finally culminating to its end state as a white dwarf.

Initially, a low-mass star, like our Sun, has a protracted period of steady hydrogen fusion at its core. This phase, known as the main sequence, supports the star throughout the bulk of its lifespan. During this moment, the outward pressure from nuclear fusion balances the inward gravitational pull, maintaining a fragile equilibrium. The star slowly transforms hydrogen into helium via the proton-proton chain reaction. This mechanism feeds the star and provides the energy required to resist gravitational collapse.

As the star matures and exhausts its hydrogen fuel, it undergoes a change. The core compresses and warms up, while the outer layers expand, causing the star to inflate into a red giant. In this phase, the star becomes a massive, brilliant object, many times bigger than its initial size. The outer layers become loosely bonded, and the star spills its material into space, generating a planetary nebula. This ejected material enriches the interstellar medium with components generated inside the star.

The ultimate step in the development of low-mass stars is the production of a white dwarf. Once the red giant loses its outer layers, the core that remains is comprised mostly of carbon and oxygen. This compact, heated core lacks the pressure support required to withstand gravitational collapse. However, it is protected from full collapse by electron degeneracy pressure—a quantum mechanical process that precludes further compression. As a consequence, the core stabilizes, and the star settles into a state of equilibrium.

In this condition, a white dwarf is exceedingly dense, with a mass equivalent to that of the Sun but squeezed into a volume around the size of Earth. It produces a faint light as it progressively cools over billions of years, finally vanishing from view. This signals the conclusion of the stellar adventure for low-mass stars, as they slowly fade away into cosmic oblivion.

In summary, the destiny of low-mass stars is a monument to the delicate interaction of gravity, nuclear fusion, and quantum physics. These stars develop from steady main-sequence luminaries to bloated red giants, shedding their outer layers in an elegant dance. The remaining core, the white dwarf, stands as a monument to the endurance of matter under the unrelenting pull of gravity. The study of these stellar relics not only reveals the history of the cosmos but also gives significant clues into its future. As low-mass stars' brilliance fades, they leave behind a legacy of enriched elements, changing the universe for ages to come.

7.2 Chandrasekhar Limit and Electron Degeneracy Pressure

The Chandrasekhar Limit is a crucial term in astrophysics, named for the Indian scientist Subrahmanyan Chandrasekhar. It describes the maximum mass that a stable white dwarf star may hold, above which it would suffer catastrophic gravitational collapse. This limit is about 1.4 times the mass of the Sun, or 1.44 solar masses to be exact. When a white dwarf's mass hits this level, the electron degeneracy pressure that maintains the star against gravitational contraction is overcome, resulting to a Type Ia supernova explosion.

Electron degeneracy pressure emerges owing to the Pauli Exclusion Principle, which asserts that no two electrons may occupy the same quantum state simultaneously. In the high-density core of a white dwarf, electrons are packed exceedingly tightly, and the pressure imposed by these electrons resists further compression. This counteracts the force of gravity attempting to collapse the star. It's vital to note that this pressure is independent of temperature and is a result of quantum physics.

Chandrasekhar's idea was groundbreaking since it gave an explanation for the process underlying Type Ia supernovae. These supernovae are vital for astrophysical processes like as nucleosynthesis and the distribution of elements in the cosmos. The explosion of a white dwarf near the Chandrasekhar

Limit unleashes a massive amount of energy, temporarily outshining whole galaxies. This surge of energy is vital in seeding the universe with heavy metals like iron, nickel, and chromium.

The Chandrasekhar Limit also has ramifications for our understanding of cosmology and the destiny of the cosmos. If the mass of a white dwarf could reach the Chandrasekhar Limit without suffering a supernova, it would lead to the development of black holes. The limit consequently imposes a limitation on the conceivable endpoints of star development.

Moreover, Chandrasekhar's work is closely intertwined with our grasp of stellar interiors and the life cycles of stars. It offers a critical boundary condition for models of stellar evolution, helping us comprehend the mechanisms that regulate the birth, life, and death of stars. It also shows the crucial significance of quantum mechanical concepts in astronomy, underlining the multidisciplinary character of current astrophysical study.

In summary, the Chandrasekhar Limit and electron degeneracy pressure stand as cornerstone principles in astrophysics. They illuminate the principles regulating the stability of white dwarf stars and give information on the catastrophic explosions known as Type Ia supernovae. Moreover, this restriction has major ramifications for our understanding of nucleosynthesis, the distribution of elements in the universe, and possibly the ultimate destiny of the cosmos. Chandrasekhar's discoveries continue to ripple across the field of astrophysics, changing our grasp of the universe at both macroscopic and microscopic dimensions.

7.3 Binary Systems and Type Ia Supernovae

Binary star systems serve a key role in the cosmos, affording a unique insight into the complexity of stellar development. A binary system consists of two stars rotating around a shared center of mass, bonded by their mutual gravitational pull. This dynamic interplay may profoundly effect the destiny of these stars. Of particular relevance is the function of binary systems in the production of Type Ia supernovae, which are crucial for understanding the expansion of the universe.

In a binary system, the stars may affect each other's development in fundamental ways. Mass transfer between the stars is a regular process, when material from one star gets accreted onto its partner. This process may lead to the regeneration of a star, perhaps postponing or modifying its final destiny. For instance, in a near binary system, a white dwarf may accrete material from its partner until it exceeds the Chandrasekhar limit, generating a Type Ia supernova. This situation is known as the single-degenerate model.

Type Ia supernovae are of essential significance in astronomy, since they serve as standard candles for measuring cosmic distances. Their extremely stable luminosities enable astronomers to quantify the immensity of the cosmos and estimate the rapidity of its expansion. Understanding the fundamental dynamics that lead to Type Ia supernovae is vital for improving our understanding of dark energy and the destiny of the cosmos.

One of the main hypotheses for Type Ia supernovae is the single-degenerate scenario. In this concept, a white dwarf accretes material from a partner star until it hits a critical mass, known as the Chandrasekhar limit. At this moment, the pressure at the core becomes so enormous that carbon fusion ignites in a runaway process, resulting to a cataclysmic explosion. The discharge of energy is astonishing, sometimes outshining an entire galaxy for a short moment. This explosion ejects vast quantities of heavy metals into space, replenishing the interstellar medium with elements needed for the development of future stars and planets.

However, the knowledge of Type Ia supernovae is not without its complications. There are alternate scenarios, such as the double-degenerate scenario, where two white dwarfs in a binary system fuse, beyond the Chandrasekhar limit and culminating in a supernova. Additionally, the specific triggering mechanism and the elements that determine the explosion's properties remain issues of continuing study.

In conclusion, binary star systems offer a dynamic background for examining the subtleties of stellar development and the catastrophic events that create our universe. Type Ia supernovae, formed by the interaction of stars in binary systems, give crucial insights regarding cosmic distances and the accelerated expansion of the universe. The single-degenerate hypothesis, including mass transfer onto a white dwarf, remains one of the primary theories for these remarkable events. Yet, continuing study continues to expand our knowledge of the mechanisms leading to Type Ia supernovae, improving our awareness of the basic forces driving the universe.

Chapter 8: Stellar Death: Supernovae and Neutron Stars

8.1 Supernova Mechanisms

Supernova processes reflect the catastrophic endpoints of stellar development, when a star suffers a spectacular explosion, releasing a huge amount of energy and frequently outshining whole galaxies for a short period. There are two basic processes that produce supernova explosions: core-collapse supernovae and type Ia supernovae. Each mechanism is connected with specific kinds of stars and underlying physical processes.

Core-collapse explosions, also known as type II supernovae, occur in big stars with initial masses exceeding roughly 8 solar masses. The core of such a star undergoes a succession of nuclear events that result in the production of iron. Unlike earlier phases of fusion, when energy is released, iron fusion consumes more energy than it creates. Consequently, the core falls under its own gravitational weight. This collapse leads to an implosion, forcing the core to decrease quickly. In a matter of moments, the inner core achieves extraordinarily high densities, and the outer layers rebound, causing a tremendous shockwave that propagates outward through the star. This shockwave fractures the star's structure, and the ejected material flashes brilliantly, providing the amazing visual spectacle typical of supernovae.

Type Ia supernovae, on the other hand, are generated by a distinct set of conditions. They occur from the eruption of a white dwarf in a binary system, when the white dwarf accretes mass from its partner star. As the white dwarf gathers mass, it approaches a critical threshold known as the Chandrasekhar limit, about 1.4 times the mass of the Sun. At this time, the degeneracy pressure sustaining the white dwarf's core becomes inadequate to oppose gravity, resulting to a fast collapse. Unlike core-collapse supernovae, type Ia explosions maintain a steady brightness, making them essential as "standard candles" for estimating cosmic distances.

In terms of their aftermath, core-collapse supernovae leave behind dense leftovers like as neutron stars or black holes. Neutron stars, exceedingly small objects formed almost entirely of neutrons, may exhibit phenomena like pulsars, generating beams of radiation as they revolve. Black holes, on the other side, exhibit such massive gravitational pressures that not even light can escape their event horizons.

Analyzing the relevance of these supernova processes shows their essential role in astrophysics. Core-collapse supernovae are vital for the dissemination of heavy atoms created in the star's core, replenishing the interstellar medium with ingredients required for the development of planets and life as we know it. Type Ia supernovae, with their steady brightness, have altered our knowledge of cosmic distances and the accelerated expansion of the universe. This, in turn, led to the formation of the notion of dark energy, one of the most fundamental findings in current cosmology.

In summary, supernova processes are among the most spectacular and scientifically impactful phenomena in the cosmos. Whether via the catastrophic collapse of huge stars or the explosive igniting of accreting white dwarfs, supernovae give windows into the severe physics driving stellar life cycles and provide critical insights into the larger universe. Their aftermath, in the form of compact relics or distributed heavy elements, determines the fate of galaxies and the possibility for life elsewhere in the cosmos.

8.2 Neutron Stars and Pulsars

Neutron stars are one of the most interesting and extreme things in the cosmos. Formed from the leftovers of enormous stars following a supernova explosion, these stars compact the mass of many suns into a sphere just roughly 20 kilometers in diameter. This extraordinarily high density arises from the collapse of the star's core, when electrons and protons combine to generate neutrons, giving to the nickname "neutron star." Due to their massive gravitational pull, neutron stars contain powerful magnetic fields, sometimes billions of times stronger than that of Earth.

One of the most remarkable phenomena related with neutron stars is the occurrence of pulsars. Pulsars are highly magnetized, revolving neutron stars that release beams of electromagnetic radiation. These beams are not random, but rather, they are aligned with the star's magnetic axis. As a pulsar spins, these beams sweep throughout space, much like the light of a lighthouse. When these beams intercept Earth's line of sight, they are detected as regular pulses of radiation, thus the word "pulsar."

The finding of pulsars in the 1960s was a momentous advance in astrophysics. Jocelyn Bell Burnell and Antony Hewish originally spotted these pulsating radio emissions, mistakenly mistaking them for messages from alien civilizations. This finding altered our knowledge of neutron stars and their extreme physical features.

Pulsars have proved to be essential instruments for astrophysics study. Their highly exact spinning cycles, which may vary from a few milliseconds to many seconds, serve as natural cosmic clocks. By monitoring the regularity of pulsar signals, scientists have been able to discover microscopic changes produced by many circumstances, such as the existence of planets or gravitational interactions with other celestial bodies. These findings have led to important breakthroughs in our understanding of general relativity and the behavior of matter under extreme circumstances.

Furthermore, the study of neutron stars and pulsars has exposed the nature of matter at nuclear densities. The extreme pressures and temperatures at a neutron star's core challenge our understanding of the basic forces that control the cosmos. Experiments involving neutron stars and pulsars add to our knowledge of the equation of state of ultra-dense matter, which has ramifications for both astrophysics and nuclear physics.

In conclusion, neutron stars and pulsars constitute some of the most interesting and perplexing things in the universe. These relics of enormous stars give a glimpse into the extremes of physics, from the inconceivable density of their cores to the mind-boggling strength of their magnetic fields. The discovery and study of pulsars have not only enhanced our knowledge of these cosmic occurrences but have also given useful instruments for a broad variety of astrophysical inquiry. As scientists continue to investigate the secrets of neutron stars and pulsars, we reveal greater insights into the underlying nature of the universe itself.

8.3 Stellar Remnants and Supernova Remnants

When enormous stars approach the conclusion of their evolutionary path, they leave behind intriguing celestial remains that give significant insights into the dynamics of the cosmos. These stellar remains appear in numerous forms, principally as white dwarfs, neutron stars, and in the most dramatic situations, black holes. Each kind suggests a varied destiny for stars based on their starting mass and other circumstances. These relics function as cosmic time capsules, recording the history of their parent stars and giving key information for understanding stellar development.

White dwarfs, the most frequent stellar leftovers, come from stars with initial masses less than around eight times that of our Sun. These very dense particles, made mostly of carbon and oxygen, exhibit a delicate balance between gravity and electron degeneracy pressure. As the core compresses and the outer layers are evacuated during a star's red giant phase, the remaining core becomes a white dwarf. These star carcasses slowly cool over billions of years, finally vanishing into cosmic oblivion. White dwarfs, albeit tiny in size, retain immense significance in the universe, since they play a critical part in events like Type Ia supernovae, which have far-reaching consequences for the study of cosmic expansion.

Neutron stars, on the other hand, are the leftovers of enormous stars that experience a core-collapse supernova explosion. With densities surpassing those of atomic nuclei, neutron stars are breathtakingly compact. They are formed nearly exclusively of neutrons, and their magnetic fields and fast rotations give birth to powerful radiation beams, detectable as pulsars. These strange objects, frequently barely a few kilometers in diameter, contradict our traditional understanding of matter. They offer useful testbeds for the behavior of matter under severe circumstances and play a crucial part in phenomena such as gamma-ray bursts, one of the most intense occurrences in the cosmos.

The most mysterious of all stellar remnants are black holes, generated when huge stars with initial masses higher than around twenty times that of the Sun succumb to gravitational collapse. These objects are so dense that even light cannot escape their gravitational force, leaving them undetectable to direct observation. We deduce their existence by their gravitational pull on neighboring matter and by the X-rays released from their accretion disks. Black holes stand as cosmic endpoints, signifying a threshold beyond which our existing knowledge of physics is pushed to its limits. They give a fascinating insight into the probable existence of places in spacetime where our ordinary physical principles fall down.

Supernova remnants, as the name indicates, are the leftovers of a giant star's spectacular dying throes. These cosmic remains are distinguished by their complicated, filamentary architecture, frequently visible in many wavelengths of light. They play a significant role in the enrichment of the interstellar medium with heavy elements, which are important for the development of successive generations of stars and planetary systems. Furthermore, the shockwaves created by supernova explosions drive fresh star formation, therefore impacting the development of galaxies.

In conclusion, stellar relics serve as witness to the dynamic and diverse nature of stellar development. From the dense, cooling embers of white dwarfs to the harsh environs of neutron stars and the cryptic enclaves of black holes, these remains provide fundamental insights into the life cycles of stars. Supernova leftovers, with their complicated architecture and far-reaching repercussions, give more windows into the cosmic fabric. Together, these remains hold the key to comprehending the basic processes that form our universe.

Chapter 9: Black Holes: The End of Massive Stars

9.1 Gravitational Collapse

Gravitational collapse is a basic astrophysical phenomenon that happens when a huge object, such as a star, approaches the end of its life cycle. This phenomena is driven by the unrelenting force of gravity, which presses downward on the star's core. As a star exhausts its nuclear fuel, the equilibrium between gravitational forces and the internal pressure caused by nuclear reactions becomes unstable. When this balance is upset, the star experiences a catastrophic metamorphosis, resulting to either the development of a compact remnant or, in the case of extraordinarily large stars, a supernova explosion.

The collapse process is triggered when a star's core can no longer hold itself against the gravitational force. For lower-mass stars like our Sun, this signifies the shift from the main sequence phase to the red giant phase. The core shrinks as the outside layers expand, generating a shell of gas encircling the center. This period may endure for millions of years, during which the star pulsates and loses its outer layers, enriching the interstellar medium with heavy elements.

As the core continues to compress, it hits a crucial stage known as the Chandrasekhar Limit. This limit, named after the Indian scientist Subrahmanyan Chandrasekhar, determines the greatest mass a non-rotating white dwarf may acquire before electron degeneracy pressure can no longer oppose gravitational forces. When a white dwarf crosses this limit, it may produce a type Ia supernova, a thermonuclear explosion that releases a huge amount of energy and ejects the bulk of the star's material into space.

In contrast, for more massive stars, the collapse process may lead to the production of neutron stars or black holes. As the core compresses, the density and temperature climb to severe extremes. Eventually, the core's electrons mix with protons to generate neutrons, resulting in a compact neutron star. If the core's mass surpasses a specific threshold, even neutron degeneracy pressure is inadequate to prevent further collapse, resulting to the development of a black hole—a area of spacetime from which nothing, not even light, can escape.

The repercussions of gravitational collapse extend well beyond individual stars. It plays a critical role in the history and dynamics of galaxies, impacting their structure and the distribution of stuff inside them. Gravitational collapse also underpins the development of planetary systems, as the collapse of gas and dust clouds gives birth to new stars and associated planetary systems.

In conclusion, gravitational collapse is a fundamental phase in the life cycle of stars, affecting their fates and altering the surrounding cosmic landscape. Whether leading to the development of neutron stars, black holes, or the magnificent explosions of a supernova, this phenomena highlights the complex interaction between gravity, nuclear physics, and the destiny of celestial entities. Its effects are felt throughout the cosmos, from the enrichment of interstellar space with heavy elements to the development of galaxies and the systems of planets that orbit them.

9.2 Event Horizons and Schwarzschild Radius

In the field of astrophysics, few ideas are as perplexing and significant as the event horizon and the

Schwarzschild radius. These two interrelated conceptions emerge from Albert Einstein's General Theory of Relativity, which transformed our understanding of gravity. At its heart, the Schwarzschild radius encompasses the crucial threshold beyond which no information, matter, or even light can escape the gravitational grasp of a huge object. This limit forms the event horizon of a black hole, a region where the very fabric of spacetime is distorted beyond recognition.

The Schwarzschild radius, named after the scientist Karl Schwarzschild, is a basic quantity associated with a non-rotating black hole. It depicts the distance from the singularity at the center of a black hole to the event horizon, beyond which escape is impossible. Mathematically defined as $R_s = 2GM/c^2$, where G is the gravitational constant, M is the mass of the black hole, and c is the speed of light, the Schwarzschild radius is a distinguishing property of black holes. For instance, a black hole with a mass of one solar mass has a Schwarzschild radius of around 3 kilometers, showing the extraordinary tightness of these cosmic phenomena.

The event horizon, on the other hand, is an imagined surface around a black hole, representing the border of no return. It delineates the zone where the escape velocity surpasses the speed of light, making any type of escape ineffective. This notion derives from the fact that General Relativity predicts the bending of spacetime around big objects. As an object approaches the Schwarzschild radius, spacetime curvature becomes so extreme that all feasible courses converge inexorably inward. This fundamental boundary, the event horizon, has important consequences for our knowledge of the cosmos.

The interaction between the Schwarzschild radius and the event horizon underlines the uncanny character of black holes. Once an item exceeds the event horizon, it becomes causally detached from the observable world. It is a point of no return, when the rules of physics as we know them break down. Paradoxically, from an external observer's viewpoint, time seems to slow down for any item entering the event horizon, finally freezing as it approaches. This effect, known as gravitational time dilation, is a witness to the enormous gravitational forces at play.

In summary, the Schwarzschild radius and the event horizon are crucial to our grasp of black holes, those cosmic enigmas that transcend conventional thinking. The Schwarzschild radius measures the point of gravitational no return, encompassing the singularity's gravitational effect. The event horizon, in turn, defines the threshold where escape from a black hole's grasp becomes impossible. Together, these principles outline the bizarre character of black holes, where spacetime itself undergoes severe contortions. They serve as testaments to the deep discoveries that General Relativity has given to mankind, changing our vision of the universe.

9.3 Active Galactic Nuclei and Quasars

Active Galactic Nuclei (AGNs) constitute some of the most energetic and intriguing phenomena in the cosmos. These are compact areas in the cores of galaxies that generate massive quantities of energy, frequently outshining the combined brightness of their host galaxies. The energy production from AGNs spans a wide range, from radio waves to gamma rays, and understanding their nature has been a primary focus of astrophysics study.

AGNs are assumed to be fueled by accretion of mass onto supermassive black holes, which may have masses millions to billions of times that of the Sun. As matter spirals inwards towards the event horizon, it generates an accretion disk, where gravitational potential energy is transformed into strong radiation over the electromagnetic spectrum. This mechanism may yield luminosities comparable to billions of stars.

One of the most severe examples of AGNs is the phenomena of quasars, which are the most energetic and bright objects in the universe. They were initially detected in the 1960s as point-like radio emitters with no apparent equivalent in visible light. Later, with developments in technology, quasars were shown to be exceedingly distant and enormously strong objects, typically situated billions of light years away. They give essential insights into the early cosmos, enabling astronomers to examine conditions and processes that occurred immediately after the Big Bang.

Quasars demonstrate a spectrum of fascinating features. Their spectra reveal significant emission lines, suggesting the existence of highly ionized gas. These lines come from multiple atomic transitions inside the accretion disk and adjacent areas. Additionally, quasars may demonstrate quick and unanticipated change in their brightness, indicating that the core engine is small and active.

The study of AGNs and quasars has far-reaching consequences for our knowledge of cosmology and the large-scale structure of the universe. Their huge luminosities enable them to be detected at cosmic distances, giving crucial probes of the early cosmos. Moreover, their involvement in galaxy formation and development is vital. AGNs are predicted to impact the interstellar medium and star formation rates in their host galaxies, hence changing the cosmic landscape on a grand scale.

However, various issues surrounding the specific processes driving AGNs and quasars remain unresolved. The accretion process onto supermassive black holes, for instance, is still not completely understood, and the physical conditions around the event horizon are impossible to see directly. Additionally, the triggering processes that lead to the activation of AGNs, as well as the interaction between

AGNs and their host galaxies, are current topics of investigation.

In conclusion, Active Galactic Nuclei and quasars constitute some of the most powerful and exciting phenomena in the universe. They are powered by the gravitational energy release from the accretion of matter onto supermassive black holes, resulting to emissions across the whole electromagnetic spectrum. Quasars, being the most luminous objects known, give a unique insight into the early cosmos. Studying AGNs not only increases our knowledge of basic astrophysical processes but also gives information on the larger cosmic setting in which galaxies and large-scale structures develop. Despite substantial advances, numerous mysteries surrounding these artifacts linger, promising intriguing discoveries in the future.

Chapter 10: Stellar Clusters and Stellar Populations

10.1 Open Clusters and Globular Clusters

Open clusters and globular clusters are two unique forms of stellar aggregation, both affording significant insights into the dynamic development of star populations inside galaxies.

Open clusters, commonly referred to as galactic clusters, are loose gatherings of very young stars that emerged from the same molecular cloud. These clusters generally consist of hundreds to thousands of stars and are gravitationally bonded, however their gravitational attraction is not strong enough to prevent members from ultimately dispersing. Due to their relatively young age, open clusters are commonly found in areas of prolific star formation, such as the spiral arms of galaxies. They serve as laboratories for researching stellar evolution, since the stars inside them share comparable ages, compositions, and distances. The Pleiades (M45) and the Hyades are noteworthy examples of open clusters inside our Milky Way.

On the other side, globular clusters are closely packed, spherical groups of stars that are substantially older than open clusters. These clusters may include tens of thousands to millions of stars, tightly clustered at their cores. They are scattered throughout the halos of galaxies, circling the galactic nucleus. Globular clusters are considered to be among of the earliest objects in the universe, going back to the early phases of galaxy formation. Their stars demonstrate a broad variety of ages, but often have extremely low metallicities, suggesting they originated when the universe's metallicity was considerably lower. Examples include Omega Centauri and M13 in the Milky Way.

Analyzing the variations between open and globular clusters gives vital insights into the origin and development of galaxies. Open clusters are linked with active star-forming areas, where molecular clouds are still extant, supplying the essential material for fresh star formation. They also give possibilities to examine the initial mass function and early star development. In contrast, globular clusters give a window into the early cosmos, offering critical insights about the environment in the early phases of galaxy formation. Their old stars serve as "fossils" of the early chemical makeup of galaxies.

Furthermore, the geographic distribution of open and globular clusters inside a galaxy offers a fascinating picture of its past. Open clusters are usually found on the galactic plane, whereas globular clusters are scattered spherically, suggesting their existence in the halo. This spatial arrangement shows that open clusters emerge inside the galactic disk, while globular clusters are likely leftovers from the early stages of galaxy formation, brought in during accretion processes.

In conclusion, open clusters and globular clusters are separate entities that give contrasting viewpoints on star populations and galaxy development. Open clusters, with their newborn stars and active star-forming areas, give insights into the continuing processes of stellar birth and early development. In contrast, globular clusters, with their old stars and distinctive chemical fingerprints, give a look into the early phases of galactic history. The study of both kinds of clusters increases our knowledge of the dynamic mechanisms forming galaxies across cosmic time.

10.2 Stellar Ages and Galactic Archaeology

Stellar ages serve as cosmic timestamps, revealing essential insights into the history and development of galaxies. This chapter dives into the exciting area of galactic archaeology, which aims to understand the chronology of stars inside a galaxy. By estimating the ages of stars, scientists may recreate the story of a galaxy's birth, its dynamic processes, and the interactions that have molded its present condition.

Galactic archaeology depends on a varied arsenal of observational methods and theoretical theories. One essential approach includes spectroscopy, where the chemical composition and brightness of stars are examined. These properties serve as crucial hints, since they fluctuate during a star's lifespan, enabling astronomers to determine its age. Additionally, the study of star clusters - groupings of stars born from the same molecular cloud - gives a picture of stellar populations at a single moment in time. By analyzing the features of these clusters, astronomers may determine the ages of its component stars, allowing a view into the former epochs of a galaxy.

Furthermore, developments in stellar evolutionary models have enhanced the accuracy of age calculations. By replicating the life cycles of stars with varied masses and compositions, astrophysicists may compare theoretical predictions with observable features. This repeated approach refines our capacity to deduce star ages, allowing for a more accurate reconstruction of a galaxy's past.

The ramifications of galactic archaeology extend well beyond the bounds of individual galaxies. It gives essential data to larger astrophysical investigations, such as understanding the dynamics of galaxy clusters, the development of the interstellar medium, and even the hunt for alien life. By grasping the ages and distributions of stars inside galaxies, astronomers may discover patterns and trends that give insight on the wider cosmic environment.

In conclusion, the study of star ages and galactic archeology provides a cornerstone of contemporary astrophysics. Through rigorous observation, theoretical modeling, and analytical approaches, astronomers may discover the complicated history of galaxies. This information not only enhances our understanding of the genesis and development of galaxies but also gives crucial insights into the greater cosmic fabric. As technology continues to progress and observational data accumulates, cosmic archaeology is set to unveil even deeper secrets about the nature and history of the universe.

10.3 The Halo, Disk, and Bulge Components

The structure of a galaxy is a complex interaction of numerous components, each contributing to its overall appearance and dynamics. Among these components, the Halo, Disk, and Bulge stand out as key factors in comprehending the architecture of a galaxy.

The Halo is the outermost area of a galaxy, characterized by a sparse population of stars, diffuse gas, and an abundance of dark matter. It goes well beyond the apparent bounds of the galactic disk and includes some of the oldest stars in the galaxy. These stars generally display low metallicity, suggesting that they originated in the early cosmos. The Halo is also significant in giving evidence for the existence of dark matter owing to its known gravitational effects on stars and gas inside the galaxy. Its dynamics have a crucial effect in the overall stability and kinematics of the galaxy.

In sharp contrast to the Halo, the Disk is the conspicuous, flattened portion of a galaxy that contains the bulk of its stars, gas, and dust. This is where active star creation predominantly occurs, giving birth to the recognizable spiral arms or other structural elements of the galaxy. The Disk is comprised of both new, hot stars and older, colder stars, showing its diversified stellar population. Additionally, it is inside the Disk that planetary systems and prospective habitable zones for exoplanets are located. The Disk's rotating motion is responsible for many of the dynamic features of a galaxy, including the observable galactic rotation curves.

At the core of many galaxies sits the Bulge, a tightly packed, spheroidal structure. It is largely comprised of older stars, generally with high metallicity, implying a history of substantial star production and enrichment. The Bulge is a critical location for the study of star populations and is connected with the existence of a supermassive black hole in the galactic core. Its existence greatly impacts the total gravitational potential of the galaxy, altering the orbits of stars and contributing to the stability of the galactic structure.

Analysing these components shows a complex tapestry of galaxy development. The Halo gives a peek into the early phases of a galaxy's development, preserving the remnants of the primordial cosmos. The Disk, on the other hand, is the dynamic engine of a galaxy, where continuing star formation and planetary system evolution affect its current condition. The Bulge, generally considered the galactic core, embodies a history of high star activity and is closely related with the occurrence of supermassive black holes.

Furthermore, the interactions between these components have far-reaching ramifications for galaxy development. Tidal interactions with adjacent galaxies, for example, may lead to the redistribution of material inside a galaxy, changing the relative proportions of the Halo, Disk, and Bulge components. Understanding the delicate interaction between these structural features is vital for deciphering the complicated life cycles and histories of galaxies throughout the universe.

Chapter 11: Binary Star Systems

11.1 Types of Binary Systems

Binary star systems, exciting celestial formations when two stars circle a similar center of mass, give a riveting view into the subtleties of stellar development. These systems provide a vast variety, covering a range of configurations that greatly alter the lives of the component stars.

Visual Binaries are defined by the direct observation of both stars, frequently detectable via telescopes. The stars' orbits may be traced throughout time, producing vital data on stellar masses, distances, and other key factors. Visual binaries serve a vital part in the calibration of the cosmic distance ladder, adding to our grasp of the immensity of the universe.

Spectroscopic Binaries give a unique challenge to astronomers. While only one star is immediately viewable, the spectral lines of both stars may be identified, showing their existence via Doppler shifts. By methodically examining these movements, astronomers may establish the orbit's shape, masses, and occasionally even the inclination of the system. Spectroscopic binaries frequently reveal unexpected events, such as eclipses or the existence of a third, invisible partner.

Eclipsing Binaries give a wonderful chance to examine stellar attributes. In these systems, the stars periodically pass in front of each other as viewed from Earth, generating a perceptible reduction in brightness. This phenomena enables for exact measurements of star radii, masses, and temperatures. Eclipsing binaries have proven essential in the development of astrophysical models and our knowledge of star evolution.

Close Binaries are defined by stars so close that they may interchange mass. This relationship greatly alters their evolutionary paths. Mass transfer, either via stellar winds or via an accretion disk, may lead to events like as novae, X-ray binaries, or even the development of exotic objects like black holes and neutron stars.

Wide Binaries indicate systems with large separations between their components. While they share a shared gravitational tie, they have substantially wider flexibility in their evolutionary pathways compared to near binaries. Wide binaries may serve as great laboratories for investigating the dynamics and chemical evolution of the interstellar medium, since their stars might have drastically differing ages and compositions.

In conclusion, binary systems, encompassing visible, spectroscopic, eclipsing, near, and broad configurations, serve as ideal laboratories for astrophysical study. They not only permit exact measurements of stellar characteristics but also give crucial insights into the numerous ways in which stars grow and interact. As we go further into the study of these dynamic duos, we reveal a better knowledge of the universe's complexity and the myriad routes that form the lives of stars.

11.2 Mass Transfer and Stellar Evolution in Binaries

Mass transfer in binary star systems is a key phenomenon that substantially impacts the development and destiny of the member stars. Binary systems, where two stars orbit around a shared center of mass, present a unique setting for studying stellar development. Mass transfer, the migration of material from one star to its partner, may occur via many methods and has varying repercussions for the developing stars. This phenomena plays a significant part in the production of exotic objects including cataclysmic variables, X-ray binaries, and even certain kinds of supernovae. In this research, we dig into the nuances of mass transfer in binary systems, its influence on the stars involved, and its larger implications for our knowledge of stellar evolution.

Mass transfer in binary systems largely happens in two unique ways: Roche-lobe overflow and wind-driven mass transfer. Roche-lobe overflow, the more prevalent process, takes occurs when a star grows beyond its Roche lobe, a teardrop-shaped zone surrounding each star within which the gravitational forces are balanced. When a star reaches its Roche lobe, it starts losing material to its partner, generally via an accretion disk created around the receiving star. This transfer of mass may drastically affect the evolutionary course of the stars. On the other side, wind-driven mass transfer includes the ejection of star material via intense stellar winds, which is subsequently trapped by the gravitational attraction of the partner. This process is often found in huge binary systems, notably those with at least one supergiant star.

The ramifications of mass transfer on the developing stars are substantial. For the star shedding mass, the process might lead to a decrease in its size, changing its evolutionary course and perhaps prolonging its lifespan. The receiving star, on the other hand, suffers a rise in mass, which might impact its core temperature, brightness, and general development. Mass transfer may also lead to the construction of accretion disks and the emission of substantial quantities of energy in the form of X-rays, making X-ray binaries a major topic of research in astronomy. Furthermore, when mass transfer happens between a white dwarf and a companion star, it may result in a type Ia supernova, a crucial source of information for astronomers researching cosmic distances and the expansion of the universe.

The study of mass transfer in binary systems has far-reaching implications for our knowledge of stellar evolution, since it offers a unique laboratory for seeing the ramifications of mass exchange on a star's structure and behavior. By examining the features of binary systems, such as the period of orbital motion, the nature of the mass transfer, and the subsequent changes in the stars' attributes, astronomers obtain insights into the fundamental mechanisms that regulate the life cycles of stars. Additionally, the study of mass transfer in binary systems enriches our knowledge of binary star formation, a widespread event in the cosmos. These results have implications beyond star astrophysics, adding to our grasp of galactic and cosmic processes, such as the chemical enrichment of galaxies and the generation of diverse kinds of supernovae.

In summary, mass transfer in binary star systems is a complicated process that effects the development and destiny of the component stars. The combination between Roche-lobe overflow and wind-driven mass transfer leads in a wide array of repercussions, affecting the paths of the stars and leading to the production of strange celestial objects. The study of mass transfer in binaries not only increases our knowledge of star development but also has larger ramifications for many fields of astrophysics, from the study of cosmic distances to the creation and enrichment of galaxies. As such, it remains an intriguing and crucial topic of inquiry in contemporary astronomy.

11.3 X-ray Binaries and Pulsar-Neutron Star Binaries

X-ray binaries are an intriguing type of binary star systems characterized by the emission of X-rays. This phenomena happens owing to the extreme gravitational interaction between a compact object—typically a neutron star or a black hole—and a companion star. As matter from the partner star accretes onto the compact object, it releases a vast amount of energy, most of it in the form of X-rays. This approach makes X-ray binaries essential instruments for examining tiny things and the surrounding environs.

One subset of X-ray binaries comprises pulsar-neutron star binaries. Pulsars are highly magnetized, revolving neutron stars that release beams of electromagnetic radiation. When one of these beams connects with Earth, it appears as regular pulses of radiation, giving origin to the term "pulsar". In a pulsar-neutron star binary, the pulsar is the compact object. As it accretes mass from its partner star, it produces X-rays, resulting in an X-ray pulsar. The study of pulsar-neutron star binaries gives insights into the behavior of neutron stars under severe settings and the mechanics of accretion.

Analyzing the importance of X-ray binaries and pulsar-neutron star binaries demands a multi-faceted approach. Firstly, they serve as vital laboratories for verifying the theory of general relativity in the strong gravity regime. The fast mobility of matter in the proximity of compact objects in these systems enables for high-precision testing of Einstein's theory, which has been validated in multiple situations. This, in turn, helps to our knowledge of the underlying nature of gravity.

Furthermore, X-ray binaries are helpful in the research of star evolution. They give a view into the last phases of a star's existence, especially in binary systems when mass transfer happens. This process may rise to occurrences like supernovae, offering crucial knowledge on the nucleosynthetic processes that create materials in the cosmos. Additionally, the study of X-ray binaries improves in our grasp of the dynamics of binary star systems, particularly the impacts of mass transfer and orbital development.

In the case of pulsar-neutron star binaries, they give unique insights into the nature of neutron stars, which are extraordinarily dense and perplexing phenomena. The high magnetic fields and quick rotation speeds of pulsars make them great natural laboratories for researching exotic states of matter, as well as the behavior of matter under severe circumstances. Additionally, the X-ray emissions from these systems allow for the calculation of the mass and radius of the neutron star, which are critical factors for understanding the equation of state of ultra-dense matter.

In conclusion, X-ray binaries and pulsar-neutron star binaries are astrophysical miracles that give a plethora of information about compact objects, gravitational physics, and stellar development. Through painstaking observations and theoretical modeling, scientists have revealed several riddles surrounding these systems. As technology progresses, additional discoveries in this subject promise to expand our knowledge of the underlying forces and processes that govern the universe.

Chapter 12: Exoplanets and Planetary Systems

12.1 The Diversity of Exoplanets

The discovery of exoplanets, planets circling stars beyond our solar system, has altered our knowledge of the universe. The enormous variety of these celestial entities challenges conventional assumptions of planetary systems. They vary from blazing hot gas giants, like the famed 51 Pegasi b, which circles its star in a scant four days, to ice-covered super-Earths, like Kepler-22b, dwelling in their star's habitable zone. Such a range challenges our previous beliefs about the circumstances essential for planetary formation and the possibility for habitability.

One of the most fascinating kinds of exoplanets are the so-called "hot Jupiters." These gas giants orbit incredibly near to their parent stars, frequently with orbits measured in days rather than years. Their presence has spurred a reevaluation of our knowledge of planetary migration. It was once considered that such big planets could not form so close to their stars owing to high temperatures limiting the creation of solid cores. The finding of hot Jupiters contradicts this idea, indicating that planetary migration could be a regular process, changing planetary systems.

On the opposite end of the spectrum are the rocky super-Earths. These planets, generally a few times the mass of Earth, may be covered in seas or have thick atmospheres, offering fascinating concerns regarding their habitability. Kepler-22b, for instance, is positioned inside its star's habitable zone, where circumstances would be hospitable to liquid water—a vital element for life as we know it. The study of super-Earths gives a look into the possible variety of conditions where life may develop.

Moreover, exoplanets have been detected in binary star systems, defying the concept that stable planetary orbits need a single star. Kepler-16b, a gas giant circling two stars, is a spectacular example. This study points to the possibility of habitable zones inside binary star systems, broadening the variety of probable conditions where life can thrive.

The study of exoplanets not only increases our knowledge of planetary formation and dynamics but also asks issues regarding the possibility for life beyond Earth. The topic of astrobiology is intricately linked with exoplanetary science, as researchers examine the circumstances essential for life's origin and survival. The sheer variety of exoplanets offers a rich canvas for these investigations, allowing a spectrum of habitats to examine.

In conclusion, the finding and research of exoplanets have opened a new frontier in astrophysics. The variety of these celestial entities contradicts our earlier ideas about planetary creation, migration, and habitability. From burning gas giants to possibly life-bearing super-Earths, exoplanets highlight the enormous spectrum of conditions that may exist in the universe. Moreover, the study of exoplanets has major implications for the hunt for alien life, as researchers probe the possible habitability of these faraway worlds. The study of exoplanets continues to be a lively and dynamic area, stretching the frontiers of our knowledge of the cosmos.

12.2 Planet Formation and Migration

Planet formation is a compelling process that occurs inside the protoplanetary disks orbiting young stars. This delicate dance of celestial bodies spans millions of years and culminates in the development of numerous planetary systems. It is a narrative of gravitational pull, collisions, and accretion, all under the watching eye of the central star.

At the foundation of this tale lies the nebular hypothesis, a notion presented by Kant and Laplace in the 18th century, which proposes that planets emerge from a spinning ring of gas and dust orbiting a newborn star. As the disk cools and condenses, minute particles cluster together, becoming planetesimals. Through reciprocal gravitational pull, these planetesimals continue to expand, ultimately giving birth to protoplanets. This process, known as accretion, leads to the development of terrestrial and gas giant planets.

However, the narrative doesn't finish here. Migration, a vital feature of planetary development, lends a dynamic element to the tale. Gravitational interactions within the protoplanetary disk, as well as with surrounding planets, may create major changes in a planet's orbit. This migration may be inward, towards the central star, or outward, towards the fringes of the system. This process has been seen both in our own solar system, with Jupiter's migration being a focus of continuing study, and in exoplanetary systems, affording insights into the complex tapestry of planetary constructions.

Analyzing this process exposes the extraordinary variety of planetary systems. Theories of planet formation and migration give insight on the origin of planetary resonances, such as those between Jupiter's moons, and the fascinating occurrence of exoplanets in tight, eccentric orbits. Moreover, the study of migration gives a key to understanding the survival and dispersion of planets in their different systems. This information is crucial in the larger effort to uncover the riddles of habitability and the possibilities for life beyond Earth.

Furthermore, the research of planet formation and migration increases our knowledge of the greater cosmic backdrop. It gives crucial insights into the creation of planetary systems in varied conditions, offering information on the possibility for life elsewhere in the cosmos. Additionally, this study enhances our knowledge of the development of star clusters and galaxies, since the mechanisms driving planet formation and migration are linked with the wider dynamics of these cosmic entities.

In conclusion, the account of planet creation and migration is a riveting chapter in the vast narrative of cosmic development. From the lowly origins of dust and gas to the amazing variety of planetary systems, this process uncovers the astounding complexity of the cosmos. The study of planet formation and migration not only increases our knowledge of our own solar system but also gives a glimpse into the awe-inspiring variety of exoplanetary systems. It stands as a tribute to the continuous curiosity of mankind in its quest to fathom the universe.

12.3 Habitable Zones and the Search for Extraterrestrial Life

The notion of habitable zones, frequently referred to as the "Goldilocks zone," is a key principle in astrobiology, underlying the hunt for alien life. This zone indicates the area surrounding a star where circumstances are suitable for liquid water to exist on the surface of a planet. Given that water is an essential element for life as we know it, the finding of habitable zones gives a beginning point in the hunt for possibly habitable exoplanets.

Within a solar system, the habitable zone is not a constant distance but changes depending on the brightness and temperature of the host star. Planets placed too near to their star will endure severe temperatures, causing water to evaporate, while those too far risk freezing over altogether. Consequently, a planet's location inside the habitable zone is a delicate balance, where circumstances are just ideal for the presence of liquid water—a necessity for life as we understand it.

Advancements in astronomy and technology have allowed scientists to identify exoplanets inside their host stars' habitable zones. Telescopes, like as the Kepler and Hubble Space Telescopes, have detected many of these possibly habitable planets. Yet, although the finding of such planets is exciting, it's vital to highlight that habitability relies on numerous conditions beyond merely being in the appropriate orbital area. Factors including atmospheric composition, geological stability, and the existence of a protective magnetic field all play essential roles in determining a planet's habitability.

Furthermore, the quest for alien life stretches beyond our own galaxy. Researchers are researching exoplanets in adjacent star systems, seeking for signals of habitability and possible biosignatures—indicators of life such as the existence of particular gasses or surface conditions. This quest is motivated by the enticing potential of finding life beyond Earth, a proposition that would alter our understanding of the cosmos and our role within it.

In recent years, missions like the James Webb Space Telescope (planned to launch) have been constructed with the main purpose of analyzing the atmospheres of exoplanets, especially those inside habitable zones. This is a tremendous leap forward in our capacity to evaluate the potential habitability of distant planets. The James Webb Space Telescope's powerful equipment will enable scientists to investigate the chemical composition of exoplanet atmospheres, offering critical insights into their potential for harboring life.

In conclusion, habitable zones constitute a fundamental idea in the hunt for alien life. They identify the zones surrounding stars where circumstances are best for liquid water, a vital element for life. Identifying planets inside these zones is a promising beginning, but actual habitability relies on a complex interaction of elements. With breakthroughs in technology and the launch of missions like the James Webb Space Telescope, we stand on the threshold of a new age in our quest to unravel the mysteries of possible life beyond our home planet. The hunt of habitable exoplanets is a tribute to humanity's persistent curiosity and our unrelenting desire to answer one of the most important questions: are we alone in the universe?

Chapter 13: Stellar Evolution in Galaxies

13.1 Galactic Dynamics and Stellar Orbits

Galactic dynamics and the migration of stars inside a galaxy comprise a basic part of astrophysics. This topic investigates the delicate dance of celestial bodies under the influence of gravitational forces inside the immense expanse of galaxies. The dynamics of these systems are regulated by Newton's principles of motion and universal gravity, establishing the basis upon which our knowledge of galaxy structures is constructed.

At its heart, the study of galactic dynamics tries to comprehend the complicated interaction between gravitational pull and the inertial motion of stars. One of the fundamental realizations in this discipline is that galaxies are not static entities but are in a state of continual motion. Stars inside a galaxy follow orbits around the galactic center, with velocities that are dictated by the total mass contained within their particular orbits. This has major consequences for our knowledge of galaxy structure, since it suggests the existence of massive quantities of dark matter, which exerts a gravitational impact on the known stars, but remains elusive to direct observation.

One of the primary observational technologies used to understand galactic dynamics is spectroscopy. By measuring the redshift or blueshift of stellar light, astronomers may estimate the line-of-sight velocities of stars. This information, when paired with comprehensive imaging, allows for the production of velocity maps inside a galaxy, showing the distribution of mass and offering insights into the underlying gravitational potential. Additionally, developments in technology have permitted the exact determination of stellar proper motions, offering a three-dimensional perspective of stellar orbits.

The study of galactic dynamics has far-reaching ramifications. It lays the framework for understanding the creation and development of galaxies, offering information on the processes that regulate the spatial distribution of stars and the construction of galactic features such as spiral arms and bars. Furthermore, it has been essential in the finding and characterisation of supermassive black holes dwelling at galactic centers. Their existence has a tremendous impact on the motion of stars in the region, frequently leading to the detection of high-velocity stars near these mysterious objects.

In conclusion, galactic dynamics and the study of star orbits form a cornerstone of contemporary astrophysics. Through rigorous observations and the application of basic physical principles, astronomers have uncovered the complicated dance of stars inside galaxies. This topic not only advances our knowledge of galaxy structures and the existence of dark matter, but also gives essential insights into the origin and development of galaxies over cosmic timescales. Furthermore, it serves as a tribute to the tremendous advances made possible by the union of theory and observation in the goal of deciphering the secrets of the cosmos.

13.2 Star Formation Histories

Star formation histories (SFHs) are key chapters in the life cycle of galaxies. They give a precise description of how and when stars were formed inside a specific galactic entity. Essentially, SFHs act as cosmic journals, charting the eons-long process of star formation. These histories are produced via astrophysical observations and simulations, enabling us to comprehend the dynamic interaction between the development of galaxies and the birth of new stars.

Astronomers utilize a range of approaches to unravel these histories. One typical technique is the examination of color-magnitude diagrams (CMDs) of stars inside a galaxy. By studying the distribution of stars in different parts of the CMD, astronomers may infer the ages and masses of stars, offering a snapshot of star creation during several epochs. Additionally, spectroscopic surveys play a key function. They let astronomers to examine the light produced by stars, disclosing their chemical composition and age. These methodologies, when performed together, present a detailed chronology of star formation.

The analysis of star formation histories frequently gives unexpected insights into galactic history. For instance, a galaxy with a sharply increasing SFH implies a powerful burst of star creation in its recent past. This might suggest an interaction or merger event, when gravitational forces induce the birth of new stars. Conversely, a galaxy with a generally constant SFH implies a more steady, quiet evolutionary course. These histories also give information on the interplay between internal processes, like as gas accretion and supernova feedback, and external impacts like interactions with nearby galaxies.

Star formation histories serve as crucial benchmarks for cosmological theories and simulations. By comparing observed SFHs with theoretical predictions, scientists may enhance our knowledge of the fundamental physics regulating the creation of stars in galaxies. This repeated procedure helps evaluate and enhance the accuracy of models, ensuring they appropriately portray the intricate interaction of gas dynamics, radiation, and gravitational forces that influence galaxy development. Furthermore, the study of star formation histories enables researchers to restrict parameters linked to the initial mass function (IMF) of stars, a basic component of stellar populations.

In conclusion, star formation histories give a fascinating view into the evolutionary narrative of galaxies. Through a mix of observational methods and theoretical modeling, astronomers may recreate the historical record of star creation inside a galaxy. This information not only helps our knowledge of galaxy development but also acts as a litmus test for cosmological theories. Ultimately, star formation histories constitute a cornerstone in the great tale of the universe, showing the delicate dance between galaxies and the stars that inhabit them.

13.3 Stellar Feedback and Galactic Evolution

Stellar feedback, a critical component of galactic development, defines the effect that stars exert on their surrounding environment via numerous mechanisms, including radiation, stellar winds, and supernova explosions. This process plays a vital role in defining the dynamics, structure, and chemical composition of galaxies across cosmic time periods. As stars originate and grow within a galaxy, they emit vast quantities of energy in the form of radiation and ejected matter. This energy adds momentum and heat to the surrounding interstellar medium (ISM), causing a series of complicated interactions that, in turn, govern star formation rates and redistribute materials necessary for the creation of new stars.

At the core of stellar feedback lies the awe-inspiring catastrophe of supernova explosions. These cataclysmic catastrophes occur when huge stars near the end of their lifespan, their cores collapsing under great gravitational strain. The subsequent explosion delivers a rush of energy, shockwaves, and a variety of heavy components into the surrounding medium. This infusion of fresh material, richer with heavy metals generated in the star's core, is important in refilling the galactic ISM with the required components for the development of following generations of stars and planetary systems.

One of the most exciting effects of stellar feedback is its potential to manage star formation inside a galaxy. The energy and momentum imparted by stellar winds and supernova explosions may compress neighboring molecular clouds, prompting the formation of new stars. However, this same feedback may also scatter and destabilize these clouds, blocking subsequent star formation. This delicate balance generates a dynamic feedback loop, as the energy released by freshly born stars impacts the production of following generations. Over prolonged time scales, this feedback loop governs the total pace at which a galaxy creates stars, playing a vital role in defining its evolutionary direction.

Moreover, star feedback adds greatly to the chemical enrichment of galaxies. Through nucleosynthesis inside star cores and supernova explosions, heavy metals are created and dispersed into the ISM. These elements, necessary for the development of planets and perhaps life, are dispersed across the cosmos. Consequently, across cosmic time spans, galaxies develop chemically, with successive generations of stars emerging from progressively metal-enriched material. This chemical development leaves an indelible stamp on the galaxy population, allowing a look into the history of star formation and nucleosynthesis within that part of the universe.

In conclusion, star feedback stands as a cornerstone of galactic development, managing a complex interaction of energy, matter, and chemical components inside a galaxy. From the catastrophic explosions of supernovae to the radiative pressures of large stars, the effect of stellar feedback reverberates across cosmic time. By controlling star formation rates, dispersing elements crucial for life, and modifying the chemical makeup of galaxies, stellar feedback determines the fate of galaxies, leaving a permanent legacy in the vast fabric of the cosmos.

Chapter 14: The Cosmic Connection: Stellar Evolution and Cosmology

14.1 The Age of the Universe

The age of the cosmos ranks as one of the most important and complex topics in cosmology. Through rigorous scientific study, astronomers have sought to answer this conundrum, pulling together hints from many parts of the universe. The consensus estimate, as of my latest knowledge update in September 2021, estimates the age of the universe at roughly 13.8 billion years. This astonishing result is based on a mix of observations, theoretical models, and empirical data derived from a range of astrophysical events.

The cornerstone of our knowledge of the universe's age is founded upon the cosmic microwave background radiation (CMB). Detected in 1965 by Arno Penzias and Robert Wilson, the CMB is a faint glow of radiation enveloping the universe, a relic of the Big Bang. This relic radiation gives a glimpse of the cosmos as it transitioned from an opaque, hot plasma to a transparent, colder state. Through painstaking observations and thorough study of the CMB, cosmologists have acquired critical information about the early universe's circumstances, enabling scientists to estimate its age with unparalleled accuracy.

Furthermore, the study of distant galaxies and their redshifts gives another key piece of the picture. Edwin Hubble's pioneering discoveries in the early 20th century discovered a universal expansion, revealing that galaxies are retreating from one another. The rate of cosmic expansion, embodied in the Hubble constant, is a vital criterion for calculating the age of the universe. By projecting backward in time, scientists arrive at a limited age for the universe.

The age of the earliest stars and stellar systems also places important restrictions on the universe's timeline. Globular clusters, ancient conglomerations of stars, host some of the oldest stellar populations known. Using advanced methods in astrophysics, researchers have dated these clusters to be almost as ancient as the universe itself, proving the consistency of the cosmic history.

However, it is vital to note that the quest of the universe's age is a dynamic and changing area of study. Advancements in technology, enhanced measuring methods, and fresh theoretical frameworks may refine our knowledge of this temporal mystery. Therefore, it is important for readers to reference updated sources to comprehend the newest estimations, particularly after September 2021.

In summary, the age of the universe includes a convergence of astrophysical evidence and theoretical models, providing an estimated age of around 13.8 billion years. This determination is based in the cosmic microwave background radiation, the expansion of the cosmos, and the ages of ancient star systems. While our present knowledge is solid, it is crucial to stay open to future discoveries that may improve or extend our perspective of the universe's temporal tapestry.

14.2 Stellar Populations and Cosmic Microwave Background

Stellar populations serve a critical role in deciphering the evolutionary history of galaxies and comprehending the greater cosmic background. A stellar population refers to a collection of stars within a galaxy that share similar properties like as age, chemical composition, and geographical distribution.

By analyzing these populations, astronomers may deduce key characteristics about a galaxy's genesis, star creation history, and its interactions with its surroundings.

One of the primary pieces of evidence supporting the Big Bang hypothesis and our knowledge of the early Universe is the cosmic microwave background (CMB) radiation. Discovered in 1965 by Arno Penzias and Robert Wilson, this faint, homogeneous light penetrates the universe and reflects the leftovers of radiation released when the Universe was barely 380,000 years old. The CMB gives a picture of the Universe's condition at that early epoch, displaying a relatively uniform and isotropic distribution of energy.

The relationship between stellar populations and the CMB originates from the fact that the features of stars in a galaxy, like as their chemical composition and evolutionary stage, are intricately tied to the galaxy's age and history. Older galaxies, for instance, are often dominated by populations of low-mass, long-lived stars, whereas younger galaxies could display zones of rapid star formation and more massive, short-lived stars. This information is crucial for understanding the observed spectral properties of galaxies over a variety of wavelengths.

The age of stellar populations may be determined from the dispersion of stars on the Hertzsprung-Russell diagram, which maps brightness versus temperature. By comparing this distribution with theoretical models of star development, astronomers may estimate the age of a specific population. This information is especially helpful for researching galaxies in the distant Universe, since it gives insights about the period of their origin.

Furthermore, the chemical makeup of stars within a population may be established using spectroscopic research. Elements heavier than hydrogen and helium, known as metals in astrophysical parlance, are created by nucleosynthesis in past generations of stars. Hence, the metallicity of stars reflects the enrichment history of their parent galaxy. This knowledge is vital in identifying the mechanisms that lead to the creation of galaxies and the development of the interstellar medium.

In summary, the study of star populations offers as a significant tool for understanding the development and history of galaxies. This information, in turn, contributes in comprehending the greater cosmic tale, including the vital insights supplied by the cosmic microwave background. By integrating measurements of stellar populations with theoretical models of cosmic evolution, astronomers may create a thorough narrative of the Universe's journey from its primordial condition to the vast array of galaxies and stars we view today.

14.3 Implications for the Fate of the Universe

The destiny of the universe stands as one of the most important issues in cosmology, integrating the history, present, and future of our cosmos. As scientists unravel the rich fabric of star development, we acquire critical insights into the fate that awaits. At the center of this research lies the interaction between expansion and gravity, a dynamic that has directed the universe's direction from its origin.

The measurements of distant galaxies, each redshifted in line with Hubble's law, give a window into the universe's expansion. This finding, initially stated by Edwin Hubble in the early 20th century, gave rise to the

Big Bang hypothesis, positing a cosmos that erupted out from a singularity. The essential aspect in deciding the destiny of the cosmos rests on the balance between the expansion's velocity and the gravitational force seeking to prevent it.

If the universe continues to expand at an accelerated pace, as current evidence shows, it will succumb to what astrophysicists name the "Big Chill" or "Heat Death." In this scenario, galaxies will become progressively separated, stars will exhaust their fuel, and the universe will get ever colder and darker. Eventually, just fragments of radiation will remain, signifying the silent end of the universe as we know it. The cosmos will turn into a wide stretch of near absolute zero, a frightening, lifeless emptiness.

Conversely, if the gravitational pull proves powerful, the cosmos may suffer a different end - the "Big Crunch." In this scenario, the expansion would ultimately end, and the universe would begin to compress. Galaxies would move close together, culminating in a catastrophic convergence. This collapse would culminate in a singularity, perhaps leading to the emergence of a new universe in a cycle of cosmic rebirth. This cyclical perspective of the world provides a sense of constant regeneration, giving a dramatic contrast to the endless darkness of the Big Chill.

Furthermore, the nature of the cosmos, notably the enigmatic dark energy and dark matter, has a major impact on its eventual fate. The cryptic nature of these creatures continues to intrigue astronomers, as their gravitational influences modify the fundamental fabric of the universe. If dark energy's effect becomes dominating, it may cause an ever-speeding expansion, accelerating the approach towards the Big Chill. Conversely, if dark matter's gravitational pull becomes more powerful, it might tilt the scales towards the ultimate reversal of expansion in a Big Crunch scenario.

In conclusion, the destiny of the cosmos, linked with the narrative of star development, is a matter of intense intrigue and wonder. The interaction of expansion and gravity, directed by the enigmatic forces of dark energy and dark matter, will eventually decide the universe' future. Whether the universe meets its end in the icy embrace of the Big Chill or enjoys a violent rebirth in a Big Crunch, the consequences for our understanding of existence itself are enormous. As we continue to peek further into the cosmos, the answers to these existential concerns may still be within our reach, affording a glimpse into the majesty and wonder of the world.

Chapter 15: Beyond Stellar Evolution: The Future of Stars and the Cosmos

15.1 Stellar Engines and Advanced Civilizations

The notion of stellar engines provides a fascinating junction of astronomy and speculative technology. It anticipates sophisticated civilizations harnessing the energy of whole stars for their purposes. The most known theoretical example is the Dyson Sphere, postulated by scientist Freeman Dyson in 1960. This megastructure would enclose a star, collecting and harnessing its energy output. While still firmly within the realm of speculative conjecture, the hypothesis invites serious issues regarding the possible capabilities of intelligent alien civilizations.

The practicality of star engines rests on the premise that civilizations might have the technical capacity to create such massive constructions. Given our present knowledge of physics and engineering, creating a Dyson Sphere is beyond our current capabilities, and it's vital to note that any society trying this would need to have past multiple technical obstacles. The creation of such a gigantic artificial structure would need sophisticated materials, energy generating methods, and a degree of social organization much beyond what we presently possess.

From an astronomical standpoint, the ramifications of star engines are similarly deep. The building of a Dyson Sphere around a star would radically change its visible features. Observing astronomers could notice odd patterns in a star's energy output, perhaps revealing the existence of a megastructure. This offers the enticing potential that astronomers may one day uncover evidence of a sophisticated civilization via their manipulation of star energy.

Furthermore, the presence of star engines has profound consequences for our understanding of the Fermi Paradox. This conundrum questions why, given the large number of possibly habitable planets in the cosmos, humans have not yet identified any signals of alien civilizations. The presence of star engines implies that sophisticated civilizations may be beyond our present technological horizon, thus explaining our lack of direct communication.

However, it's vital to approach the notion of stellar engines with a degree of care. While it's an intriguing thinking experiment, it remains firmly stuck in the domain of speculative conjecture. We have no substantial proof of any sophisticated society, much alone one capable of erecting such megastructures. Moreover, there are several scientific, technical, and logistical obstacles connected with establishing a Dyson Sphere or equivalent star engine.

In conclusion, the notion of stellar engines provides an intriguing marriage of astrophysical theory and speculative technology. It sees sophisticated civilizations harnessing the energy of whole stars for their purposes, most famously illustrated by the Dyson Sphere. While the notion is now beyond our technical capabilities, it invites significant considerations about the possible possibilities of intelligent alien civilizations. From an astronomical standpoint, the identification of star engines might alter our knowledge of the universe. However, it's vital to approach this notion with care, acknowledging it as a theoretical construct that presently sits firmly in the domain of scientific conjecture.

15.2 The Fate of the Last Stars

In the big cosmic drama, as eons stretch into infinity, the destiny of the final stars unfurls with a sad inevitability. These star remains, formed in the cores of ancient galaxies, now stand as the climax of a cosmic drama that has endured billions of years. At this terminal stage, two primary archetypes emerge: white dwarfs and neutron stars. Each contains its own unique tale of the universe's twilight.

The white dwarf, a remnant of lower-mass stars, arises from the ashes of a dying sun-like star. Held together by the relentless pressure of electron degeneracy, these ethereal remains resist gravity collapse, becoming a cosmic reminder of once dazzling cores. Over time, however, even these strong sentinels bow to the inexorable march of entropy. Slowly, over the millennia, they will decay, becoming cold, black cinders floating into the wide expanse of space. White dwarfs, in their serene twilight, reflect the silent epilogue to the magnificent acts of their ancestors.

In sharp contrast, neutron stars, the leftovers of gigantic stellar infernos, reflect a cosmic transformation of unthinkable extremes. These intriguing remains occur from supernova explosions, when the enormous forces released during a star's last convulsions compress its core into a tight nucleus of neutrons. The resultant entity, astoundingly small but packed with vast mass, pulsates with magnetic forces capable of distorting the very fabric of spacetime. Pulsars, the lighthouses of the cosmos, are formed in this catastrophic dance, unleashing beams of radiation that sweep over the universe in regular cadence.

As the eons tumble ahead, even neutron stars succumb to the inescapable tug of time. Through mechanisms still hidden in mystery, they may finally bow to the inexorable pull of gravity, compressing deeper into the ultimate enigma: a black hole. Within the cosmic singularity, existing principles of physics approach their limits, and the very fabric of space and time contorts beyond human knowledge. In this last act, the remaining remains of stars, once beacons of brightness and energy, depart from the cosmic stage, leaving only an emptiness veiled in mystery.

In pondering over the destiny of the final stars, we are faced with the fleeting nature of all things in the universe. From the blazing birth in ancient nebulae to the dying whispers of light, stars embody the ever-turning wheel of creation. Their progress and final extinction serve as a sobering reminder of the impermanence that underpins even the most lasting celestial occurrences. Yet, in this transience, there lurks a beauty and depth that resonates through the annals of cosmic history, beckoning contemplation and wonder from those who dare to look upon the night sky.

15.3 Cosmic Implications of Stellar Evolution

Stellar evolution, the process by which stars undergo tremendous changes during their lives, holds huge cosmic ramifications. This process not only impacts the form and dynamics of galaxies but also effects the availability of components vital for life as we know it. As stars develop, they produce heavier elements via nucleosynthesis, eventually leading to their cataclysmic deaths in supernovae. This release of energy and newly created elements has far-reaching repercussions for the interstellar medium and the possibility for planetary formation.

The nucleosynthesis happening inside stars throughout their development carries a profound cosmic meaning. Elements heavier than hydrogen and helium, such as carbon, oxygen, and iron, are formed in the cores of stars by nuclear fusion. When large stars go supernova, they discharge these newly produced atoms into space, enriching the interstellar medium with a wide palette of chemical components. This process has direct ramifications for the development of following generations of stars and planets. It creates a chemical legacy that impacts the makeup of subsequent star systems, giving the basic elements essential for the formation of life.

Furthermore, the violent deaths of big stars in supernovae have a tremendous cosmic influence. Supernova explosions generate tremendous quantities of energy, momentarily outshining whole galaxies. These occurrences may create shockwaves that compress adjacent gas clouds, spurring the formation of new stars. Additionally, the immense heat and energy from supernovae may influence the dynamics of neighboring galaxies, impacting their structure and development. In this manner, the life and death of stars play a fundamental part in changing the cosmic landscape.

One of the most exciting cosmic implications of star evolution concerns to the formation of life-supporting planets. The heavier elements generated in the cores of stars, as well as those expelled during supernova explosions, create the building blocks of planets. Without the input of stars, the development of rocky planets with the essential chemical variety to sustain life would be unlikely. Our own existence, therefore, owes greatly to the complicated processes of star development.

In a larger cosmic framework, understanding star development gives essential insights into the age and destiny of the universe. By examining the numbers of stars in galaxies and the chemical makeup of stellar remains, astronomers may impose restrictions on the age of galaxies and the universe as a whole. Additionally, the study of stellar evolution influences our knowledge of the ultimate destiny of the universe, since it throws insight on processes such as stellar remnants (e.g., white dwarfs, neutron stars, and black holes) and their interactions with the surrounding environment.

In conclusion, the study of star development exposes enormous cosmic implications that resound throughout a broad spectrum of astrophysical phenomena. From nucleosynthesis to the development of planets and galaxies, stars act as cosmic laboratories, altering the physical and chemical composition of the universe. Moreover, their life cycles give vital insights into the larger cosmic environment, including the age and final destiny of the universe. Stellar evolution serves as a witness to the interdependence of celestial processes and their effect on the magnificent fabric of the universe.

Conclusion

In the voyage through the complexity of stellar evolution, we have uncovered the intriguing narrative of how stars, the celestial engines that form the universe, grow during their cosmic lifespan. From their modest origins as dense cores amid molecular clouds, stars undergo a transforming process, ending in a magnificent display of energy and, ultimately, their destiny. This voyage through the chapters has led us from the formation of stars in protostellar clouds, through the main sequence, the grandeur of red giants, and the catastrophic end points of supernovae and black holes.

One of the most significant disclosures comes in the discovery of stellar nucleosynthesis, the alchemical process by which stars make elements, including the basic building blocks of life, via the fusing of lighter atoms. This process, happening inside the heart of stars, engenders the birth of heavy elements, distributing them throughout the universe following stellar death, therefore seeding future generations of stars and planets. This cosmic recycling serves as a striking reminder of the interdependence of all celestial entities.

The ramifications of stellar evolution extend well beyond the bounds of individual stars. Stellar clusters and populations give crucial insights into the dynamic history of galaxies. By monitoring these pockets of star activity, astronomers may identify the age and composition of galaxies, unraveling the great tale of the universe's development. Moreover, binary star systems provide a unique view into the intricate interplay between celestial partners, offering information on the varied spectrum of interactions that create the stellar landscape.

As we explore farther into the worlds of exoplanets and planetary systems, we unearth the possible cradles of life beyond our own light blue dot. The endeavor to locate habitable zones and seek for alien lifeforms offers a beacon of hope and amazement in our study of the universe. The results from exoplanet investigations not only increase our grasp of the cosmos, but also excite the imagination with the enticing potential of life living somewhere in the wide expanse of space.

In pondering the destiny of stars and the universe, we encounter both the awe-inspiring and the sobering. The relentless march of time and the inexorable force of entropy define the eventual fate of stars. We see into the distant future, imagining a universe progressively cooling and disappearing, with the final stars flickering out like embers in a cosmic fire. Yet, inside this cosmic tapestry, lies the promise for inconceivable futures, as evolved civilizations harness the energy of stars and negotiate the obstacles offered by the developing cosmos.

In conclusion, the adventure through star development is a monument to the immense beauty and complexity of the universe. From the birth pains of protostars to the catastrophic finales of supernovae, the life cycles of stars affect the fundamental fabric of the universe. Each step of star development unravels a new layer of knowledge, and in doing so, improves our appreciation for the beauties that exist in the cosmic universe. The history of stars stretches well beyond their physical life, weaving a tale that transcends epochs and galaxies, leaving an everlasting stamp on the universe itself.

Refrences

1. Adams, F. C., & Laughlin, G. (1997). A dying universe: The long-term fate and evolution of astrophysical objects. Reviews of Modern Physics, 69(2), 337.

2. Arnett, W. D. (1996). Supernovae and nucleosynthesis: An investigation of the history of matter, from the Big Bang to the present. Princeton University Press.

3. Carroll, B. W., & Ostlie, D. A. (2017). An introduction to modern astrophysics. Cambridge University Press.

4. Chabrier, G., & Baraffe, I. (2000). Theory of low-mass stars and substellar objects. Annual Review of Astronomy and Astrophysics, 38(1), 337-377.

5. Clayton, D. D. (1968). Principles of Stellar Evolution and Nucleosynthesis. McGraw-Hill.

6. Cox, A. N., & Giuli, R. T. (2008). Principles of stellar structure. Cambridge University Press.

7. Dwek, E. (1998). The galactic cycle of carbon through stars, interstellar medium, and interstellar dust. The Astrophysical Journal, 501(2), 643.

8. Frank, J., King, A., & Raine, D. J. (2002). Accretion power in astrophysics. Cambridge University Press.

9. Freedman, W. L., & Kaufmann, W. J. (2007). Universe. Macmillan.

10. Heger, A., & Woosley, S. E. (2002). The nucleosynthetic signature of population III. The Astrophysical Journal, 567(1), 532.

11. Kippenhahn, R., & Weigert, A. (1990). Stellar structure and evolution. Springer-Verlag.

12. Maeder, A. (2009). Physics, formation and evolution of rotating stars. Springer Science & Business Media.

13. Mihalas, D., & Binney, J. (1981). Galactic astronomy: Structure and kinematics. W. H. Freeman.

14. Nomoto, K., Thielemann, F. K., & Yokoi, K. (1984). Explosive nucleosynthesis in carbon deflagration models for type I supernovae. The Astrophysical Journal, 286, 644.

15. Ostriker, J. P., & Hartwick, F. D. A. (1968). Spiral structure in galaxies: A density wave theory. The Astrophysical Journal, 153, 797.

16. Paczynski, B. (1971). Evolution of single stars. Annual Review of Astronomy and Astrophysics, 9(1), 183-208.

17. Reimers, D. (1975). Circumstellar envelopes and mass loss of red giant stars. Memoirs of the Société Royale des Sciences de Liège, 8, 369.

18. Salaris, M., & Cassisi, S. (2005). Evolution of stars and stellar populations. John Wiley & Sons.

19. Schaller, G., Schaerer, D., Meynet, G., & Maeder, A. (1992). New grids of stellar models from 0.8 to 120 solar masses at Z= 0.020 and Z= 0.001. Astronomy and Astrophysics Supplement Series, 96, 269-331.

20. Schwarzschild, M., & Härm, R. (1959). On the evolution of stars of various masses. Astrophysical Journal, 129, 637.

21. Stahler, S. W., & Palla, F. (2004). The formation of stars. Wiley-VCH.

22. Trimble, V., & Aschwanden, M. J. (2006). Solar system astrophysics: Planetary atmospheres and the outer solar system. W.B. Saunders.

23. Trimble, V., & Reipurth, B. (2013). Encyclopaedia of Astronomy and Astrophysics. Nature Publishing Group.

24. Van Belle, G. T. (2012). Binary stars: A decade of discoveries. Nature, 491(7423), 8.

25. Weaver, T. A., Zimmerman, G. B., & Woosley, S. E. (1978). The evolution and explosion of massive stars. Astrophysical Journal Supplement Series, 219, 937.

26. Weinberg, S. (1972). Gravitation and cosmology: Principles and applications of the general theory of relativity. John Wiley & Sons.

27. Wolszczan, A., & Frail, D. A. (1992). A planetary system around the millisecond pulsar PSR1257+12. Nature, 355(6356), 145.

28. Woosley, S. E., & Weaver, T. A. (1995). The evolution and explosion of massive stars. II. Explosive hydrodynamics and nucleosynthesis. The Astrophysical Journal Supplement Series, 101, 181.

29. Yoon, S. C., & Langer, N. (2005). Evolution of rapidly rotating metal-poor massive stars towards gamma-ray bursts. Astronomy and Astrophysics, 443(2), 643.

30. Zeilik, M., Gregory, S. A., & Smith, E. V. P. (2012). Introductory Astronomy & Astrophysics. Saunders College Publishing.

31. Zhang, Q., Fall, S. M., & Whitmore, B. C. (2001). Star formation in the interacting galaxies NGC 5752 and NGC 5753. The Astrophysical Journal, 561(2), 727.

32. Zwicky, F. (1933). Die Rotverschiebung von extragalaktischen Nebeln. Helvetica Physica Acta, 6, 110.

33. Zwicky, F. (1937). On the masses of nebulae and of clusters of nebulae. The Astrophysical Journal, 86, 217.

34. Zwicky, F. (1942). The redshift of extragalactic nebulae. The Astrophysical Journal, 95, 329.

35. Zwicky, F. (1957). Morphological astronomy. Springer.

36. Zwicky, F. (1961). First steps in numerical astrophysics. Berlin, Springer.

37. Zwicky, F. (1963). Selected papers on the galaxy and the cosmos. University of Chicago Press.